OPTICAL AND ELECTRICAL PROPERTIES

PHYSICS AND CHEMISTRY OF MATERIALS WITH LAYERED STRUCTURES

Managing Editor

E. MOOSER, *Laboratoire de Physique Appliquée, CH - 1003, Lausanne, Switzerland*

Advisory Board

E. J. ARLMAN, *Bussum, The Netherlands*

F. BASSANI, *Physics Institute of the University of Rome, Italy*

J. L. BREBNER, *Department of Physics, University of Montreal, Montreal, Canada*

F. JELLINEK, *Chemische Laboratoria der Rijksuniversiteit, Groningen, The Netherlands*

R. NITSCHE, *Kristallographisches Institut der Universität Freiburg, West Germany*

A. D. YOFFE, *Department of Physics, University of Cambridge, Cambridge, U.K.*

VOLUME 4

OPTICAL AND ELECTRICAL PROPERTIES

Edited by

P. A. LEE

Brighton Polytechnic, England

D. REIDEL PUBLISHING COMPANY

DORDRECHT-HOLLAND/BOSTON-U.S.A.

Library of Congress Cataloging in Publication Data

Main entry under title:

Optical and electrical properties.

(Physics and chemistry of materials with layered structures; v. 4)
Includes bibliographies and indexes.
1. Solids—Optical properties. 2. Solids—Electric properties. 3. Layer structure (Solids). I. Lee, Peter A., 1926-   II. Series.
QD478.P47 vol. 4 [QC176.8.06]   530.4'1s   [530.4'1]
ISBN 90-277-0676-X   76-40337

Published by D. Reidel Publishing Company
P.O. Box 17, Dordrecht, Holland

Sold and distributed in the U.S.A., Canada, and Mexico
by D. Reidel Publishing Company, Inc.
Lincoln Building, 160 Old Derby Street, Hingham,
Mass. 02043, U.S.A.

All Rights Reserved
Copyright © 1976 by D. Reidel Publishing Company, Dordrecht, Holland
No part of the material protected by this copyright notice may be reproduced or
utilized in any form or by any means electronic or mechanical,
including photocopying, recording or by any informational storage and
retrieval system, without written permission from the copyright owner

Printed in The Netherlands

TABLE OF CONTENTS

PREFACE	VII
B. L. EVANS / Optical Properties of Layer Compounds	1
J. BORDAS / Some Aspects of Modulation Spectroscopy in Layer Materials	145
R. ZALLEN and D. F. BLOSSEY / The Optical Properties, Electronic Structure and Photoconductivity of Arsenic Chalcogenide Layer Crystals	231
P. M. WILLIAMS / Photoemission Studies of Materials with Layered Structures	273
R. C. FIVAZ and PH. E. SCHMID / Transport Properties of Layered Semiconductors	343
D. J. HUNTLEY and R. F. FRINDT / Transport Properties of Layered Structure Metals	385
R. F. FRINDT and D. J. HUNTLEY / Experimental Aspects of Superconductivity in Layered Structures	403
J. M. VANDENBERG-VOORHOEVE / Structural and Magnetic Properties of Layered Chalcogenides of the Transition Elements	423
INDEX OF NAMES	459
INDEX OF SUBJECTS	461

PREFACE

This fourth volume in the series 'Physics and Chemistry of Materials with Layered Structures' is concerned with providing a critical review of the significant optical and electrical properties by established authors who have themselves made many significant contributions to these fields.

Research into these materials has recently gained a new impetus and their fascinating properties have attracted many new research workers. These people should find much of value in the reviews contained in this volume and the editor is very much indebted for the painstaking and hard work put into the preparation of the various chapters by the authors.

The optical properties provide useful information for deriving the band structures, a knowledge of which is required for an interpretation of measurements on the electronic properties. The chapters by Dr Evans, Dr Williams and Dr Bordas describe different techniques which have provided much detailed data on this subject. An interesting property of these materials is the comparative ease with which thin specimens may be prepared for these measurements and this is highlighted in the superconducting experiments outlined by Professor Frindt and Dr Huntley. These authors together with Dr Vandenberg's chapter on the magnetic properties also describe the interesting and significant intercalation mechanisms whereby a wide range of organic compounds and alkali metals may be incorporated in the lattice. This provides an additional parameter for varying the properties of these materials and may yet be seen to provide eventual possible applications of layer compounds.

The arsenic chalcogenides have been extensively studied for their photoconductive properties and Dr Zallen and Dr Blossey have written a useful and extensive review of this subject. These particular materials have analogous amorphous or glassy structures and a comparison is made with this equally rapidly growing field of research.

The transport properties have been reviewed from both the semiconducting point of view (Dr Fivaz and Dr Schmid) and the metallic properties (Dr Huntley and Professor Frindt). These show quite clearly the increasing understanding we now have of the electrical properties of these materials.

In the preparation of a book of this kind there is inevitably a time lag before publication and in what has now become a rapidly expanding field of research much new data is continually becoming available. However, the extensive data provided by the various authors should give the necessary platform and information source for workers in this field.

I would like to thank the authors for their co-operation and forbearance in the preparation of this volume, and to Professor E. Mooser, the General Editor of this series of volumes, for his encouragement and invaluable advice. I am also indebted to Dr A. B. Yoffe for his helpful suggestions and discussions in the initial composition of this volume.

Criticisms due to shortcomings or omissions in the preparation of this volume should be levelled at the editor alone, but with recent developments in this rapidly expanding field there is much scope for further contributions to future volumes in this series.

Brighton Polytechnic, England DR P. A. LEE

OPTICAL PROPERTIES OF LAYER COMPOUNDS

B. L. EVANS

Physics Department, University of Reading, England

1.	INTRODUCTION	2
2.	INTERBAND ABSORPTION THEORY	3
	2.1 Direct Interband Transitions	3
	2.2 Indirect Interband Transitions	13
3.	EXCITON THEORY OF ABSORPTION	14
	3.1 Delocalized (Wannier) Excitons	15
	3.2 Delocalized (M_0) Excitons: Optical Selection Rules	20
	3.3 Delocalized Hyperbolic (M_1, M_2) Excitons	23
	3.4 Exciton Effects at an M_3 Critical Point	25
	3.5 Intermediate Excitons	25
	3.6 Indirect Exciton Transitions	26
	3.7 Exciton Line Broadening	27
4.	FREE CARRIER ABSORPTION	28
	4.1 Classical Model	29
	4.2 Electron Energy Band Model	29
5.	THE EFFECT OF AN APPLIED ELECTRIC FIELD ON THE CRYSTAL DIELECTRIC FUNCTION	30
	5.1 Effective Mass Approximation (EMA)	30
	5.2 Forbidden Interband Transitions	36
	5.3 Indirect Transitions	37
	5.4 Excitonic Transitions	38
	5.5 Symmetry Analysis of Electro Reflectance Spectra	41
6.	THE EFFECT OF AN APPLIED MAGNETIC FIELD ON THE CRYSTAL DIELECTRIC FUNCTION	41
	6.1 Simple Energy Bands	42
	6.2 Complex Energy Bands	43
	6.3 Delocalized Excitons in a Magnetic Field	43
	6.3.1 Parabolic (M_0) Excitons	43
	6.3.2 Hyperbolic Excitons	45
	6.4 Intra and Interband Magneto Absorption in Semiconductors	47
	6.5 Exciton Absorption in a Magnetic Field	49
	6.5.1 Parabolic Excitons	49
	6.5.2 Hyperbolic Excitons	51
7.	COMBINED ELECTRIC AND MAGNETIC FIELDS	51
	7.1 Semiconductor Having Simple Energy Bands	51
	7.2 Semiconductors Having Complex Energy Bands	52
	7.3 Indirect Transitions	52
	7.4 Exciton Transitions	53
8.	STRESS MODULATED SPECTRA	
9.	THE EFFECT OF TEMPERATURE ON THE CRYSTAL DIELECTRIC FUNCTION	54
	9.1 Temperature Modulated Indirect Transition Spectra	54
	9.2 Temperature Modulated Direct Transition Spectra	55
	9.3 Temperature Modulated Plasma Resonance Spectra	56

10.	THE MEASURED OPTICAL PROPERTIES OF LAYER COMPOUNDS	58
	10.1 Group II Dihalides	58
	10.2 Group IV Halides	64
	10.3 Group V Halides	70
	10.4 Transition Metal Halides	73
	10.5 Group III Chalcogenides	76
	10.6 Group IV Dichalcogenides	87
	10.7 Group V Chalcogenides	90
	10.8 Transition Metal Dichalcogenides	103
	10.9 Graphite	131
REFERENCES		134
ACKNOWLEDGEMENTS		143

1. Introduction

This chapter is primarily concerned with the ways in which optical absorption/reflection measurements can give information about the electron energy band structure of layer crystals. Infra-red measurements of the atomic vibration spectra are given at the end of the chapter.

Layer crystals, as the name implies, are built up from crystalline layers stacked in a regularly repeating sequence. The layer thickness, defining a unit cell dimension, is typically 10 Å whereas the lateral dimension of the layer are those of the crystal. Two limiting cases can be envisaged. (1) The bonding between the layers is comparable with the intra layer bonding resulting in an anisotropic 'three dimensional' crystal. (2) The interlayer bonding is comparatively weak, in this case the properties of the crystal are those of the individual layer. In practice the interlayer bonding is such that the two dimensional model is never strictly valid although some features of this model have been employed to describe the properties of a few types of layer crystal.

One important consequence of the layer like nature is that many of these crystals show interlayer cleavage. As a result it is possible, by repeated cleaving using transparent adhesive tape, to prepare very thin crystals on which optical transmission measurements can be made even in the intrinsic absorption region. For the same reason these cleaved crystal surfaces are optically flat with the result that the normal incidence transmissivity and reflectivity of a crystal slice have the theoretical values predicted from the measured bulk crystal dielectric constant [1]. An exception occurs in the case of very thin crystals where surface effects modify the crystal dielectric constant [2].

In principle the one-electron energy band structure of a crystal is established by matching the calculated joint density of states, J_{cv}, with the measured absorption spectrum of the crystal. Singularities in the joint density of states at critical points, see Section 2, give rise to a characteristic structure which, when identified in the measured absorption spectrum, allow the specific critical point transition energies to be determined. In practice some degree of electron-hole interaction exists which not only introduces an additional exciton absorption but also modifies the interband absorption structure, Section 3, so that direct identification of the relevant critical point transition energy is impossible. Further information about the critical point type is gained from electric, magnetic and stress field measurements, Sections 4, 5, 6 and 7.

2. Interband Absorption Theory

This resumé of intrinsic absorption processes begins with the expression [3] for the transition probability per unit time w_{ml} if the absorbed incident radiation is monochromatic and transitions can occur to any of a group of closely spaced or continuously distributed final states m

$$w_{ml} = \frac{2\pi}{\hbar} \rho(E_m) |H'_{ml}|^2 \quad (1)$$

$\rho(E_m)$ is the density of final states (energies approx. E_m) H'_{ml}, as derived by first order perturbation theory, is given by

$$H'_{ml} = \frac{ie\hbar}{mc} \int \psi_m^* \exp(i\mathbf{q}\cdot\mathbf{r}) \mathbf{A}\cdot\nabla\psi_l \, d\tau \quad (2)$$

where

$$\mathbf{A}(\mathbf{r}, t) = \mathbf{A}_0 \exp[i(\mathbf{q}\cdot\mathbf{r} - \omega t)] + cc \quad (3)$$

is the vector potential representation of the electromagnetic field and ψ_l, ψ_m are the initial and final state wave functions. The initial state may be a discrete state or one of a continuous range of states.

2.1. Direct Interband Transitions

Equations (1–3) can be applied to semiconductors and insulators having a full valence band (v.b) and empty conduction band (c.b) since such a system conforms to the requirement that there is a continuous range of final (empty c.b) states available.

The wave function $\psi_{v\mathbf{k}_1}(\mathbf{r})$ of a v.b state can be written

$$\psi_{v\mathbf{k}_1}(\mathbf{r}) = N^{-(1/2)} e^{i\mathbf{k}_1\cdot\mathbf{r}} u_{v\mathbf{k}_1}(\mathbf{r}) \quad (4)$$

where N is the number of unit cells in crystal volume V and $u_{v\mathbf{k}_1}(\mathbf{r})$ is a Bloch function. Substituting (4) and a similar expression for the c.b wave function $\psi_{c\mathbf{k}_2}(\mathbf{r})$ in (2) gives

$$H'_{cv} = \frac{ie\hbar}{Nmc} A_0 \sum_j e^{i(\mathbf{k}_1 - \mathbf{k}_2 + \mathbf{q})\cdot\mathbf{R}_j} \int_{\text{cell}} \phi(\mathbf{r}, \mathbf{k}_1, \mathbf{k}_2, \mathbf{q}) \, d\tau \quad (5)$$

where, because of the cell periodicity of $u_{v\mathbf{k}_1}$, $u_{c\mathbf{k}_2}$ the integral has been replaced by a sum over the N unit cells of the crystal, $\mathbf{R}_j$ is a lattice vector determining the jth cell. The summation in (5) is zero unless $\mathbf{k}_1 - \mathbf{k}_2 + \mathbf{q} = \mathbf{K}$, a reciprocal lattice vector, or, on the reduced zone scheme

$$\mathbf{k}_1 - \mathbf{k}_2 + \mathbf{q} = 0. \quad (6)$$

The radiation wave vector $\mathbf{q}$ is normally negligibly small compared with the reduced wave vectors $\mathbf{k}_1$, $\mathbf{k}_2$ which are of the order of the reciprocal lattice spacing. Condition (6) therefore becomes

$$\mathbf{k}_1 \simeq \mathbf{k}_2 \tag{7}$$

so that H'_{cv} is non-zero for direct 'vertical' transitions in the reduced zone scheme.

Writing $\mathbf{k}_1 = \mathbf{k}_2$ (5) becomes

$$H'_{cv} = \frac{ie\hbar}{mc} A_0 \int_{\text{cell}} [u^*_{c\mathbf{k}_1}\boldsymbol{\alpha}\text{ grad }u_{v\mathbf{k}_1} + i\boldsymbol{\alpha}\mathbf{k}_1 u^*_{c\mathbf{k}_1} u_{v\mathbf{k}_1}] d\tau \tag{8}$$

where $\boldsymbol{\alpha}$ is a unit vector in the $\mathbf{E}$ direction, i.e. $\mathbf{A}_0 = A_0\boldsymbol{\alpha}$.

Unless the first term in the integrand of (8) is very small, as in forbidden transitions, the second term is negligible by comparison so that (8) reduces to

$$H'_{cv} = \frac{e}{mc} A_0 \boldsymbol{\alpha} \cdot \mathbf{p}_{cv} \tag{9}$$

where

$$\mathbf{p}_{cv} = -i\hbar \int_{\text{cell}} u^*_{c\mathbf{k}_1} \text{ grad } u_{v\mathbf{k}_1} d\tau. \tag{10}$$

Substituting (9) in (1) gives the number of valence to conduction band transitions occurring per unit time per unit crystal volume

$$w_{cv} = \frac{2\pi}{\hbar} |A_0|^2 \left(\frac{e}{mc}\right)^2 J_{cv} |\boldsymbol{\alpha} \cdot \mathbf{p}_{cv}|^2 \tag{11}$$

where J_{cv} is the joint density of states

$$J_{cv} = \frac{1}{4\pi^3} \int_S \frac{ds}{|\nabla_k(E_c - E_v)|_{E_c - E_v = \hbar\omega}} \tag{12}$$

ds being an element of surface in $\mathbf{k}$ space defined by $E_c(\mathbf{k}) - E_v(\mathbf{k}) = \hbar\omega$. Dividing w_{cv} by the photon flux density gives the crystal absorption coefficient α_{cv}.

$$\alpha_{cv} = \frac{4\pi^2 e^2}{n\omega m^2 c} J_{cv} |\boldsymbol{\alpha} \cdot \mathbf{p}_{cv}|^2. \tag{13}$$

ε_2, the imaginary part of the crystal dielectric constant, is given by $\varepsilon_2 = nc\alpha/\omega$ where n is the real part of the refractive index. Structure in $\alpha_{cv}(\omega)$ is mainly due to the frequency dependence of J_{cv} since $|\boldsymbol{\alpha} \cdot \mathbf{p}_{cv}|^2$ varies only slowly with frequency.

The analytic character of J_{cv}, 5(12), and hence $\varepsilon_2(\omega)$ is singular at those frequencies [4] for which

$$\nabla_k E_c(\mathbf{k}) - \nabla_k E_v(\mathbf{k}) = 0 \tag{14}$$

i.e. at critical points on the surface $\hbar\omega = E_c - E_v$ where the slope of the c.b is equal to that of the v.b. Critical points are classified into two types [5] viz

(a) Symmetry interband types for which

$$\nabla_k E_c(\mathbf{k}) = \nabla_k E_v(\mathbf{k}) = 0. \tag{15a}$$

(b) General interband types for which

$$\nabla_k E_c(\mathbf{k}) = \nabla_k E_v(\mathbf{k}) \neq 0. \tag{15b}$$

Let $\mathbf{k}_0$ define the critical point at which $|\nabla_k E_{cv}(\mathbf{k})|_{k=k_0} = 0$ where $E_{cv}(\mathbf{k})$ has been written for $E_c(\mathbf{k}) - E_v(\mathbf{k})$. At a nearby point $\mathbf{k}$ in the Brillouin zone $E_{cv}(\mathbf{k})$ can be written in the expanded form [4].

$$E_{cv}(\mathbf{k}) = E_{cv}(\mathbf{k}_0) + \sum_{i=1}^{3} a_i (k_i - k_{0i})^2. \tag{16}$$

The critical point at $\mathbf{k}_0$ is labelled M_t where t is the number ($t = 0, 1, 2, 3$) of negative coefficients in (16). Thus M_0 corresponds to a minimum in $E_{cv}(\mathbf{k})$ at k_0, M_3 to a maximum and M_1, M_2 to saddle points.

The coefficients a_i in (16) represent the relative interband curvatures; in effective mass notation

$$a_i \propto (1/m_i^c) - (1/m_i^v) \tag{17}$$

where c, v refer to conduction, valence states. In the case of large effective masses a_i is small, a limiting two dimensional case occurs when one of the coefficients a_i becomes vanishingly small.

In terms of the angular frequency ω (16) becomes

$$\omega_{cv}(\mathbf{k}) - \omega_{cv}(\mathbf{k}_0) = a(k_x - k_{0x})^2 + b(k_y - k_{0y})^2 + c(k_z - k_{0z})^2. \tag{18}$$

Transforming the origin of coordinates in $\mathbf{k}$ space to $\mathbf{k}_0$ and writing $\Omega = \omega_{cv}(\mathbf{k}) - \omega_{cv}(\mathbf{k}_0)$ (18) becomes

$$\Omega = ak_x^2 + bk_y^2 + ck_z^2. \tag{19}$$

The behaviour of $J_{cv}(\omega)$ near a critical point is best determined from the behaviour of the integrated density $g(\Omega)/4\pi^3$ [6] which gives the number of level differences lying below the energy $\hbar\omega$. Generally $g(\Omega)$ is the volume of $\mathbf{k}$ space bounded by $\Omega = ak_x^2 + bk_y^2 + ck_z^2$ and, possibly, the Brillouin zone (BZ) planes. Referred to k_0 the BZ planes can be replaced by the isotropic cut off limits $\pm K$, this over simplifies the geometry but provides the correct Ω dependence near $\Omega = 0$. The general expression for $g(\Omega)$ is

$$g(\Omega) = 8c^{-(1/2)} \int_0^{\lambda_1} dk_x \int_0^{\lambda_2} dk_y (\Omega - ak_x^2 - bk_y^2)^{1/2} \tag{20}$$

where limits λ_1, λ_2 and the signs of a, b determine the following four cases.

Case I.
An M_0 critical point a, b, $c > 0$. If $a \to 0$ this becomes an $M_0(2D)$ two dimensional critical point.

For $\Omega > 0$ energy surface (19) is an ellipsoid which, it is assumed, lies wholly within the BZ in the y, z directions. Then $\lambda_2 = b^{-(1/2)}(\Omega - ak_x^2)^{1/2}$ and λ_1 is either $(\Omega/a)^{1/2}$ or, in the $2D$ limit as $a \to 0$, simply K.

For $\Omega < 0$, $g(\Omega) = 0$ so that

$$M_0; \; g_0(\Omega) = \begin{cases} 0 & \text{when } \Omega < 0 \\ \dfrac{4\pi}{3}(abc)^{-(1/2)}\Omega^{3/2} & \text{when } o < \Omega < aK^2 \end{cases} \quad (21)$$

$$M_0(2D); \; \lim_{a \to 0} g_0(\Omega) = \begin{cases} 0 & \text{when } \Omega < 0 \\ 2\pi(bc)^{-(1/2)} K\Omega & \text{when } \Omega \gg aK^2 \end{cases} \quad (22)$$

Case II. a, b, $c < 0$ i.e. M_3, M_2 $(2D)$.
It follows from (21, 22) that when $\Omega > 0$ then $g_3(\Omega) = g_2(\Omega) = 0$ but for $\Omega < 0$ $g_3(\Omega) = g_0(-\Omega)$ and $g_2(\Omega) = g_0(-\Omega)$.

Case III. $a \le 0$; $b, c > 0$ i.e. M_1, $M_1(2D)$. For $\Omega > 0$ the energy surface is a hyperboloid of one sheet, while for $\Omega < 0$ it has two sheets. In both cases the surfaces are truncated by the BZ planes i.e. $\lambda_1 = K$. Provided $|\Omega| \ll aK^2$, bK^2 then for $\Omega < 0$,

$$g_1(\Omega) = 2\pi(bc)^{-(1/2)} \int_{(\Omega/a)^{1/2}}^{K} dk_x (\Omega + |a| k_x^2) \quad (23a)$$

and for $\Omega > 0$

$$g_1(\Omega) = 2\pi(bc)^{-(1/2)} \int_0^K dk_x (\Omega + |a| k_x^2). \quad (23b)$$

Hence at an M_1 critical point

$$g_1(\Omega) = 2\pi(bc)^{-(1/2)} \begin{cases} \Omega K + \tfrac{1}{3}|a| K^3 + \tfrac{2}{3}\left|\dfrac{\Omega^3}{a}\right|^{1/2}, & \Omega < 0 \\ \Omega K + \tfrac{1}{3}|a| K^3 & \Omega > 0. \end{cases} \quad (24)$$

The two dimensional limit is approached by allowing either a negative or positive mass to become large. When $|a| \to 0$ then, for $|\Omega| \ll bK^2$ $\lim_{a \to 0} g_1(\Omega) = \lim g_0(\Omega)$. For the case $b \to 0$ it is first necessary to consider the condition $\Omega > bK^2$, then λ_2 must be replaced by K so that

$$g_1(\Omega) = 8c^{-(1/2)} \int_0^K dk_x \int_0^K dk_y (\Omega + |a| k_x^2 - bk_y^2)^{1/2}. \quad (25a)$$

The $M_1(2D)$ case is then obtained by making the scale change $b^{1/2}k_y = k'_y$ and passing to the limit $b \to 0$; thus

$$\lim_{b \to 0} g_1(\Omega) = \begin{cases} 8c^{-(1/2)}K \displaystyle\int_{(\Omega/a)^{1/2}}^{K} dk_x(\Omega + |a|k_x^2)^{1/2}, & \Omega < 0 \\ 8c^{-(1/2)} \displaystyle\int_0^K dk_x(\Omega + |a|k_x^2)^{1/2}, & \Omega > 0. \end{cases} \quad (25b)$$

Making the substitutions $k_x = \frac{1}{2}|\Omega/a|^{1/2}(t \mp t^{-1})$ for $\Omega \gtrless 0$ shows that the dominant contribution near $\Omega = 0$ (coming from t^{-1}) is given by

$$M_1(2D); \lim_{b \to 0} g_1(\Omega) = 4|ac|^{-(1/2)} \Omega \ln 2K \left|\frac{a}{\Omega}\right|^{1/2}. \quad (26)$$

Case IV. $a \ge 0$; b, $c \le 0$. i.e. M_2, $M_2(2D)$.
For this case $g_2(\Omega) = g_1(-\Omega)$ in three dimensions.

Summary
The frequency dependence of $J_{cv}(\Omega)$, $(= dg/d\Omega)$, at the specific types of critical point (c.p), (21–26), are shown graphically in Figure 1, three dimensional c.p's, and Figure 2, two dimensional c.p's.

Assuming $|\boldsymbol{\alpha} \cdot \mathbf{p}_{cv}|$, (13), is constant over the frequency range of the critical point transitions then the frequency dependence of $\varepsilon_2(\omega)$ is that of $J_{cv}(\omega)$. In this case singularities of the type shown in Figure 1(a, b, c, d) give rise to 'edges' in the

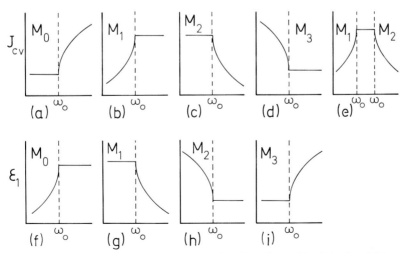

Fig. 1(a–e). Frequency dependence of J_{cv} (also ε_2 for allowed transitions) in the vicinity of three dimensional critical points; (f–i) frequency dependence of $\varepsilon_1(\omega)$ for allowed direct transitions near three dimensional critical points. [6]

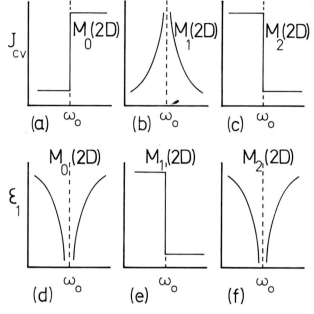

Fig. 2(a–c). Frequency dependence of J_{cv}, also ε_2 for allowed transitions, in the vicinity of two dimensional critical points; (d–f) frequency dependence of $\varepsilon_1(\omega)$ for allowed direct transitions near two dimensional critical points. [6]

$\varepsilon_2(\omega)$ spectrum and the accidental near degeneracy of M_1 and M_2 singularities, Figure 1(e), gives rise to an absorption peak. When the frequency dependence of $\varepsilon_2(\omega)$ is the same as $dg/d\omega$ then $\varepsilon_1(\omega)$ can be deduced by analytic continuation; the form of $\varepsilon_1(\omega)$ at the various c.p's is given in Figures 1(f, g, h, i) and Figures 2(d, e, f).

The expressions for $\varepsilon_2(\omega)$, $\varepsilon_1(\omega)$ can be combined to give $\hat{\varepsilon}(\omega)$ ($=\varepsilon_1(\omega)+i\varepsilon_2(\omega)$) as a function of $\xi=(\omega-\omega_0)/\omega$ for the different M types viz.

$$M_t(3D); \quad \hat{\varepsilon}(\omega) = A + Ci^{t+1}\xi^{1/2} \tag{27}$$

$$M_0(2D); \quad \hat{\varepsilon}(\omega) = A + C\ln-\xi \tag{28a}$$

$$M_1(2D); \quad \hat{\varepsilon}(\omega) = A - iC\ln\xi \tag{28b}$$

$$M_2(2D); \quad \hat{\varepsilon}(\omega) = A + C\ln\xi \tag{28c}$$

Equations (27, 28) were derived for allowed transitions where the c.b and v.b wave functions are such that at the c.p $\boldsymbol{\alpha}\cdot\mathbf{p}_{cv}(\mathbf{k}_0)\neq 0$ and is a constant or only slowly varying function of ω. Expanding $\boldsymbol{\alpha}\cdot\mathbf{p}_{cv}(\mathbf{k})$ to first order in $\mathbf{k}$ leads to the general form [6]

$$\boldsymbol{\alpha}\cdot\mathbf{p}_{cv}(\mathbf{k}) = \boldsymbol{\alpha}\cdot\mathbf{p}(0) + p_1\alpha_x k_x + p_2\alpha_y k_y + p_3\alpha_z k_z \tag{29}$$

where coefficients p_i depend upon the site $\mathbf{k}_0$. If transitions are forbidden at $\mathbf{k}_0$ (i.e. $\boldsymbol{\alpha}\cdot\mathbf{p}(0)=0$) then $|\boldsymbol{\alpha}\cdot\mathbf{p}_{cv}(\mathbf{k})|^2$ is given by the square of the remaining expression (29) which, in evaluating $\varepsilon_2(\omega)$, must be included in the integration over the same region as for allowed transitions. Confining the sum to the immediate neighbourhood of $\mathbf{k}_0$ means that cross terms $k_x k_y$ etc. will average out leaving a weighting factor dependent upon the mean polarization viz.

$$p_1^2 \alpha_x^2 k_x^2 + p_2^2 \alpha_y^2 k_y^2 + p_3^2 \alpha_z^2 k_z^2.$$

If the x, y, z axes are distinguishable then the light polarization cases $\mathbf{E} \| \mathbf{x}$, or $\mathbf{y}$ or $\mathbf{z}$ must be treated separately. Thus for the x polarization ($\mathbf{E} \| \mathbf{x}$) the contribution to $\varepsilon_2(\omega)$ is given by $p_1^2 \alpha_x^2 (dI/d\Omega)$ where

$$I_x(\Omega) = 8c^{-(1/2)} \int_0^{\lambda_1} k_x^2 \, dk_x \int_0^{\lambda_2} dk_y (\Omega - ak_x^2 - bk_y^2)^{1/2}. \tag{30}$$

This can be evaluated, as in (20), for the various limits λ_1, λ_2 and signs of 'a' and 'b'. Similarly for the y and z polarizations. In this way expressions for $\varepsilon_2(\omega)$ for forbidden (3 polarizations) transitions at $M_t(3D)$, $M_0(2D)$, $M_1(2D)$ and $M_2(2D)$ c.p's can be derived together with the corresponding expressions for $\varepsilon_1(\omega)$. These can be combined to give $\hat{\varepsilon}(\omega)$ as a function of $\xi = (\omega - \omega_0)/\omega_0$ as follows:

$$M_t(3D); \quad \hat{\varepsilon}(\omega) = A + Ci^{t+1} \xi^{3/2} \tag{31}$$

for x, y and z polarizations, in the immediate vicinity of $\mathbf{k}_0$

$$M_0(2D); \begin{cases} \hat{\varepsilon}(\omega) = A + C\xi \ln -\xi, & x, z \text{ polarizations} \quad (32a) \\ \hat{\varepsilon}(\omega) = A + C \ln -\xi, & y \text{ polarization} \quad (32b) \end{cases}$$

$$M_1(2D); \begin{cases} \hat{\varepsilon}(\omega) = A + iC[\xi \ln \xi + \xi/2], & x \text{ polarization} \quad (33a) \\ \hat{\varepsilon}(\omega) = A - iC \ln \xi & y \text{ polarization} \quad (33b) \\ \hat{\varepsilon}(\omega) = A - iC[\xi \ln \xi + \xi/2], & z \text{ polarization} \quad (33c) \end{cases}$$

$$M_2(2D) \begin{cases} \hat{\varepsilon}(\omega) = A - C\xi \ln \xi, & x \text{ polarization} \quad (34a) \\ \hat{\varepsilon}(\omega) = A + C \ln \xi, & y \text{ polarization} \quad (34b) \\ \hat{\varepsilon}(\omega) = A - C\xi \ln \xi, & z \text{ polarization} \quad (34c) \end{cases}$$

The labelling of the axes is arbitrary. For the (2D) c.p's the equations for $\hat{\varepsilon}(\omega)$, for polarizations in the infinite mass direction, correspond to those for allowed (all polarizations) transitions (28). The forms of $\varepsilon_2(\omega)$ in the vicinity of the (3D) c.p's are shown in Figure 3 and in the vicinity of (2D) c.p's in Figure 4.

Comparing Figures 1 and 2 with Figures 3 and 4 it is evident that absorption 'edges' due to forbidden transitions are less distinct than those for allowed transitions. Consequently at frequencies above the absorption threshold features in the $\varepsilon_2(\omega)$ spectrum due to forbidden transitions tend to be obscured by edges due to allowed transitions at neighbouring frequencies.

Equations (27, 28) and (31, 32, 33, 34) give the form of the complex dielectric function $\hat{\varepsilon}(\omega)$ for allowed and forbidden interband transitions at all M-type c.p's.

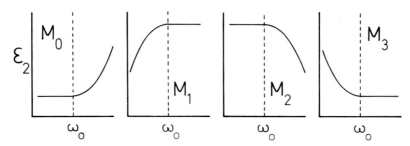

Fig. 3. The form of $\varepsilon_2(\omega)$ due to forbidden transitions at three dimensional critical points. Near ω_0 the form of $\varepsilon_2(\omega)$ is the same for all light polarizations i.e. $E \parallel x$, y and z.

In principle the nature of the c.p associated with a particular transition can be determined by matching measured $\hat{\varepsilon}(\omega)$ with one of (27, 28, 31–34). Experimentally however the measurement of $\hat{\varepsilon}(\omega)$ is a complicated process involving the separate determination of the refractive index n and absorption index $\varkappa$. For this reason it would be much more useful if c.p identification could be made directly from the measured transmissivity T or reflectivity R spectra. Using the interference free formulae $T = (1 - R^2) \exp(-\omega \varkappa l / c)$ and $R = (n + i\varkappa - 1)/(n + i\varkappa + 1)$ the

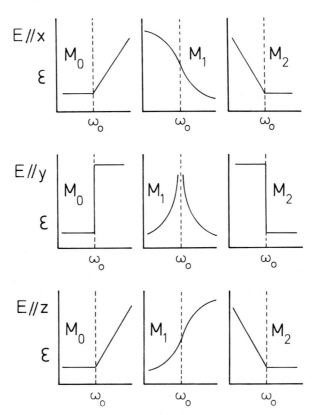

Fig. 4. The form of $\varepsilon_2(\omega)$ due to forbidden transitions at two dimensional critical points for light polarized $\mathbf{E} \parallel \mathbf{x}$, $\mathbf{E} \parallel \mathbf{y}$, $\mathbf{E} \parallel \mathbf{z}$.

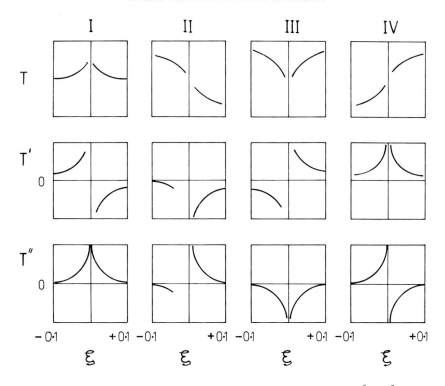

Fig. 5. The four characteristic traces of transmission, T, and derivatives $dT/d\omega$, $d^2T/d\omega^2$ as a function of $\xi = \omega_0^{-1}(\omega - \omega_0)$ for a selected set of parameters [6].

T and R spectra associated with each of the $\hat{\varepsilon}(\omega)$ spectra (27, 28, 31–34) have been calculated over the ξ range -0.1 to 0.1 for various values of parameters A and C. It was found that the T traces for *all* M_t, $M_t(2D)$ could be grouped into the four types I, II, III, IV shown in Figure 5 which also shows the associated derivative traces $dT/d\omega$, $d^2T/d\omega^2$. No other type of figure occurred even for the most extreme parametric values. Similarly the R traces grouped into the types shown in Figure 6 which also shows the associated derivatives traces.

It is in the assignment of trace type to M type that the most important results emerge. Thus by suitably varying parameters A and C it was found that the M_0 function gave either a type I or type II trace but not III or IV. Similarly $M_0(2D)$ could give types I or III. This type of behaviour occurs for all M types, both allowed and forbidden and is summarized in Tables I and II. The $3D$ forbidden transitions, which are weaker than the allowed, are hardly discernible in the direct T or R traces but show up in the second derivatives T'', R''.

In the least informative case a type III T trace could be due to either of the $3D$ M_1, M_2 types or any of three $2D$ types. If, however, the allowed transitions can be discounted for symmetry reasons then the choice is limited to the $2D$ M_0, M_1, M_2 types appropriate to the large positive mass direction; this could possibly be checked by varying the polarization direction of the incident light.

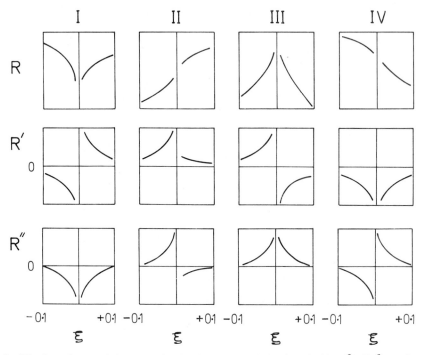

Fig. 6. The four characteristic traces of reflection, R, and derivatives $dR/d\omega$, $d^2R/d\omega^2$ as a function of $\xi = \omega_0^{-1}(\omega - \omega_0)$ for a selected set of parameters [6].

TABLE I

Assignment of three-dimensional M-types to the four characteristic transmission and reflection traces

Trace	Three dimensional M-types	
I	M_0	M_3
II	M_0	M_1
III	M_1	M_2
IV	M_3	M_3

TABLE II

Assignment of two dimensional M-types to the four characteristic transmission and reflection traces

| | Two-dimensional M-types | | | |
| | Allowed | | Forbidden | |
Trace	(all polarizations)	(x polarization)	(y polarization)	(z polarization)
I	M_0 M_2		M_0 M_2	
II		M_0 M_2		M_0 M_1 M_2
III	M_0 M_1 M_2		M_0 M_1 M_2	
IV		M_0 M_1 M_2		M_0 M_2

It follows from Figures 5 and 6 and Tables I and II that direct or derivative transmission (reflection) spectra do not uniquely identify the singular type involved. At best they provide alternatives from which a choice must be made on symmetry or other grounds.

For extreme choice of parameters A and C, covering the range $1 < n < 4$, $0 < \varkappa < 2.5$, it was sometimes difficult to distinguish between direct T traces I and IV. The ambiguity was always resolved however by comparison of the derivative traces T', T'' which provide faithful characters. In this case therefore frequency modulation could be decisive in identifying a trace and hence possible M type. Also in the case of type II or IV traces the direct T or R traces give a poorer indication of ω_0 than do the corresponding derivative traces although the geometrical centre of say T' or T'' for type II is perhaps 10% displaced from true ω_0.

2.2. Indirect Interband Transitions

In cases where the v.b and c.b extrema do not occur at the same $\mathbf{k}$ value transitions across the intervening energy gap E'_g are accompanied by the simultaneous absorption or emission of a phonon energy E_p. The absorption coefficient due to these allowed second-order processes is given by [7, 8]

$$\alpha = C(\hbar\omega - E_p - E'_g)^2 + D(\hbar\omega + E_p - E'_g)^2. \tag{35}$$

The weak absorption (~ 10 cm^{-1}) due to these indirect transitions is only visible at the absorption threshold where it is not masked by the strong absorption due to neighbouring direct transitions.

At low temperatures, where only phonon emission is possible, $D = 0$, (35), so that absorption commences at a photon energy $\hbar\omega = E_p + E'_g$ and increases approximately as the square of the excess energy, there being as many absorption components as there are phonon energies E_p. At higher temperatures $D \neq 0$, the absorption commences at the lower photon energy $\hbar\omega = E'_g - E_p$ and moves to progressively lower energies with increasing temperature as the available phonon energies increases.

In a two dimensional layer structure the potential in which the charge carriers move can be expressed as the sum of two parts, one dependent on z, the coordinate normal to the layers, the other dependent on x and y. The electron wave functions are then written as

$$\psi_{p q \mathbf{k}}(x, y, z) = S_{p\mathbf{k}}(x, y) Z_q(z) \tag{36}$$

where the S functions are solutions of a Schrödinger equation containing that part of the potential periodic in x and y while the Z_q are wave functions of an electron in a one-dimensional potential well. In this case [9] the absorption due to allowed indirect transitions is given by

$$\alpha_{\text{all}} \propto (\hbar\omega - E'_g \pm E_p) \tag{37a}$$

and for forbidden indirect transitions by

$$\alpha_{\text{forb}} \propto (\hbar\omega - E'_g \pm E_p)^2. \tag{37b}$$

3. Exciton Theory of Absorption

In addition to the structure due to interband transitions the intrinsic absorption spectrum of non-metallic crystals often contains well defined absorption peaks. This is particularly evident at the absorption threshold which seldom has the featureless forms described in Section 2 and where absorption occurs at energies less than E_g or $E'_g - E_p$. This discrepancy between theory and experiment is due to the simplicity of the 'independent particle' model which neglects the electron-hole interaction.

Crystals in which electron-hole interaction occurs are classified into three types.

Type I
In this class of solid the atoms interact only weakly with one another. An excited atomic state will therefore resemble that of an isolated atom and only extend as far as the nearest neighbour atoms say. This excitation, which represents an excited state of the entire crystal, consists of a closely coupled electron-hole pair known as a localized (or Frenkel) exciton. The Frenkel model of an exciton has been extensively employed in calculations of the excited states of molecular crystals [10, 11].

Type II
Crystals in which there is strong atomic interaction so that the excited state no longer resembles that of the isolated atom. In this class of solids (which includes semiconductors) the weakly bound exciton encompasses a number of atomic diameters and is known as a delocalized (or Wannier) exciton. When the electron-hole separation r is large compared with the lattice parameters then the intervening atomic potentials are 'averaged out' and, for many purposes, the delocalized exciton can be regarded as an electron-hole pair embedded in a dielectric. The potential energy of this pair is $-e^2/\varepsilon r$ where ε is a dielectric constant representing the screening of the intervening atoms. It is easy to see on this model how, when the exciton binding energy becomes very small, the system resembles the independent particle model of 2.1.

Type III
In this intermediate class of solids the binding energy of the 1s parabolic exciton is comparable with the band gap and, in general, neither the Frenkel nor Wannier theories completely describe the complicated spectra of such solids. One approach to the problem of excitonic effects in this intermediate class of solids is based on a modified electron-hole interaction potential. The simplest of these models uses a short range Koster-Slater [12] interaction potential which is non-zero only when the Wannier electron and hole are in the same unit cell.

The optical properties of layer compounds in zero and applied fields have been described in terms of the theories appropriate to Type II and Type III solids. The relevant properties of Wannier and intermediate excitons are as follows.

3.1. DELOCALIZED (WANNIER) EXCITONS

The energy levels in a Type II crystal where electron-hole interaction gives rise to Wannier [13] excitons can be derived from [14] the two particle model described earlier. On this model the electron-hole interaction is assumed to be $-e^2/\varepsilon |\mathbf{r}|$ and the periodic potential due to the, otherwise ignored, atoms of the crystal gives the electron and hole particles (say) *isotropic* effective masses of m_e^*, m_h^* respectively. The Schrödinger equation of the two-particle system is

$$\left(\frac{-\hbar^2}{2m_e^*}\nabla_e^2 - \frac{\hbar^2}{2m_h^*}\nabla_h^2 - \frac{e^2}{\varepsilon r}\right)\psi = E\psi. \tag{38}$$

Making the centre of mass transformation

$$\mathbf{r} = \mathbf{r}_e - \mathbf{r}_h \tag{39a}$$

$$\mathbf{R} = (m_e^*\mathbf{r}_e + m_h^*\mathbf{r}_h)/(m_e^* + m_h^*) \tag{39b}$$

(38) becomes

$$\left(-\frac{\hbar^2}{2\mu}\nabla_\mathbf{r}^2 - \frac{\hbar^2}{2M}\nabla_R^2 - \frac{e^2}{\varepsilon r}\right)\psi(\mathbf{r},\mathbf{R}) = E\psi(\mathbf{r},\mathbf{R}) \tag{40}$$

where

$$1/\mu = 1/m_e^* + 1/m_h^* \tag{41a}$$

$$M = m_e^* + m_h^*. \tag{41b}$$

The eigenfunctions can be taken of the form

$$\psi(\mathbf{r},\mathbf{R}) = e^{i\mathbf{K}\cdot\mathbf{R}}\psi(\mathbf{r}). \tag{42}$$

Making this substitution in (40) and separating it is found that $\psi(\mathbf{r})$ must satisfy the simpler equation

$$\left(-\frac{\hbar^2}{2\mu}\nabla_\mathbf{r}^2 - \frac{e^2}{\varepsilon r}\right)\psi = \left(E - \frac{\hbar^2 K^2}{2M}\right)\psi. \tag{43}$$

The solutions of (43) are those of a hydrogen atom of reduced mass μ and electronic charge $e/\sqrt{\varepsilon}$. The conclusion therefore is that for each value of K there exists a set of bound states at energies

$$E_n(\mathbf{K}) = -\frac{\mu e^4}{2\hbar^2\varepsilon^2 n^2} + \frac{\hbar^2 K^2}{2M}, \tag{44}$$

where $n = 1, 2, 3, \ldots$ is a quantum number and the last term is the kinetic energy of the bound pair. For every hydrogenic state there is a large number of states labelled by different $\mathbf{K}$ vectors. In a periodic crystal potential $\mathbf{K}$ will range over the first Brillouin zone and an excition band results.

The zero of energy in (44) is that of the state in which the exciton is dissociated to give a free electron and positive hole; on the independent particle model this state corresponds to the bottom of the conduction band. If the zero of energy is chosen as the ground state energy of the crystal then (44) becomes

$$E_n(\mathbf{K}) = E_g - \frac{\mu e^4}{2\hbar^2 \varepsilon^2 n^2} + \frac{\hbar^2 K^2}{2M}. \tag{45}$$

When the electron-hole separation is sufficiently great that the angular rotation frequency of the exciton $\omega(=\hbar n/2\mu r_e^2)$ is less than the restrahlen frequency ω_R of the crystal then the dielectric constant ε in (45) is the static dielectric constant ε_0. When $\omega > \omega_R$ i.e. $r_e^2 < \hbar n/2\mu\omega_R$ then ε is the high frequency dielectric constant ε_∞[15]. Thus for an $n=1$ exciton say $\varepsilon = \varepsilon_\infty$ while for $n=2, 3, 4$ etc $\varepsilon = \varepsilon_0$ resulting in a non-hydrogenic energy spacing of the first two levels.

A further modification to (45) occurs if there is a concentration $\bar{n}$ of conduction electrons and free holes present in the crystal. In this case, because of free charge screening, the electron-hole interaction potential becomes

$$V(\mathbf{r}) = -\frac{e^2}{\varepsilon r}\exp\left(-\frac{r}{r_d}\right). \tag{46}$$

It can be shown [16, 17] that bound exciton states will not occur if

$$r_d \leq \frac{\varepsilon \hbar^2}{\mu e^2}. \tag{47}$$

If the screening length r_d is given by the Debye expression

$$r_d^2 = \frac{\varepsilon kT}{4\pi \bar{n} e^2} \tag{48}$$

then, depending upon the values of μ, ε for the material, a certain minimum carrier concentration $\bar{n}$ will result in the disappearance of the bound exciton states.

Equations (38–45) were derived assuming isotropic effective masses m_e^*, m_h^*. In an anisotropic (hexagonal) crystal the principal dielectric constant along the c (optic) axis direction is $\varepsilon_\parallel$ while in the perpendicular (xy) plane it is $\varepsilon_\perp$. With the same axes the (positive) electron, hole masses are $m_e^\parallel$, $m_e^\perp$ and $m_h^\parallel$, $m_h^\perp$ respectively. In such an anisotropic medium the electric potential $V(\mathbf{r})$ at $\mathbf{r}$ due to a charge e at the origin is given by

$$V(\mathbf{r}) = e(\varepsilon_\parallel \varepsilon_\perp)^{-(1/2)}\left(x^2 + y^2 + \frac{\varepsilon_\perp}{\varepsilon_\parallel} z^2\right)^{-(1/2)} \tag{49}$$

In terms of the centre of mass coordinates x, y, z and making the scale change $(\varepsilon_\perp/\varepsilon_\parallel)^{1/2} z = z'$ the two-particle Schrodinger equation can be written as [18]

$$\left\{\frac{-\hbar^2}{2\mu_\perp}\left(\frac{\partial^2}{\partial x^2} + \frac{\partial^2}{\partial y^2}\right) - \frac{\hbar^2}{2\mu_\parallel}\left(\frac{\varepsilon_\perp}{\varepsilon_\parallel}\frac{\partial^2}{\partial z'^2}\right)\right.$$
$$\left. - e^2(\varepsilon_\parallel \varepsilon_\perp)^{1/2}(x^2 + y^2 + z'^2)^{-(1/2)}\right\}\psi = E\psi, \tag{50a}$$

where

$$\frac{1}{\mu_\perp} = \frac{1}{m_e^\perp} + \frac{1}{m_h^\perp} \quad \text{and} \quad \frac{1}{\mu_\parallel} = \frac{1}{m_e^\parallel} + \frac{1}{m_h^\parallel}. \tag{50b}$$

Equation (50) is conveniently rewritten as [19]

$$\left\{ \frac{-\hbar^2}{2\mu_0} \nabla^2 - \frac{1}{3} \frac{\gamma}{\mu_0} \frac{\hbar^2}{2} \left(\frac{\partial^2}{\partial x^2} + \frac{\partial^2}{\partial y^2} - \frac{2\partial^2}{\partial z'^2} \right) \right.$$

$$\left. - \frac{e^2}{\varepsilon_0} (x^2 + y^2 + z'^2)^{-(1/2)} \right\} \psi = E\psi, \tag{51a}$$

where

$$\frac{1}{\mu_0} = \frac{1}{3} \left(\frac{2}{\mu_\perp} + \frac{\varepsilon_\perp}{\varepsilon_\parallel \mu_\parallel} \right); \quad \varepsilon_0 = (\varepsilon_\parallel \varepsilon_\perp)^{1/2} \tag{51b}$$

and the anisotropy parameter γ is defined by

$$\frac{\gamma}{\mu_0} = \frac{1}{\mu_\perp} - \frac{\varepsilon_\perp}{\varepsilon_\parallel \mu_\parallel}. \tag{51c}$$

The special case $\gamma = 0$ corresponds, (51c), to the scaling condition

$$\varepsilon_\perp \mu_\perp = \varepsilon_\parallel \mu_\parallel \tag{52a}$$

(51a) then reduces to the spherically symmetric form and the eigenvalues E (measured from the bottom of the conduction band) are given by

$$E_n = -\frac{\mu_\perp e^4}{\varepsilon_0^2 2\hbar^2 n^2}, \quad n = 1, 2 \ldots \tag{52b}$$

and the exciton radii by

$$r_n = \frac{e^2 n^2}{2chR_e \varepsilon}, \tag{52c}$$

where R_e is obtained from (52b). Substituting, refer to (49), ε_0 or $\varepsilon_\perp$ for ε in (52c) gives $r_\perp$ or $r_\parallel$ the exciton radius normal ($\perp$) or parallel ($\parallel$) to the crystal c axis. If $\varepsilon_\parallel < \varepsilon_\perp$ then $r_\perp > r_\parallel$ so that the excitons are flattened in the c direction. It follows from (52b) that provided (52a) is satisfied then anisotropy, however extreme, leaves the hydrogenic energies unchanged, although the wave function may be highly distorted.

Consider the case $\gamma \neq 0$. The physically likely values of γ range from zero to the limiting two dimensional value of $\frac{3}{2}$ (see later). The operator $\Gamma \equiv (\partial^2/\partial x^2 + \partial^2/\partial y^2 - 2\partial^2/\partial z'^2)$ in (51a) has zero expectation value for s-states and connects hydrogenic states n, l, m for which $\Delta l = \pm 2, 0$. Solutions of (51a) as a function of γ ($0 \leq \gamma \leq 1$) have been determined [20] for $n \leq 4$ and are shown in Figure 7. Note that the l, m degeneracies are removed but the energies belonging to (200) and (300) are only slightly lowered over the range $0 < \gamma < 1$. Thus for 'allowed transitions', occurring between the crystal ground state and exciton s-states, only small departures from the n^{-2} spectral law are expected even for

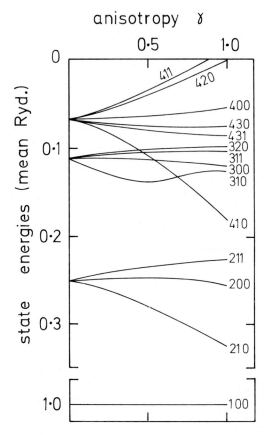

Fig. 7. Shift in exciton state energies (mean Rydberg units $e^4\mu/2\hbar^2\varepsilon^2$) due to anisotropy perturbation [20].

substantial non-scaling anisotropy. Thus for $0<\gamma<1$ and s-state excitons

$$E_n \simeq -\frac{\mu_0 e^4}{2\hbar^2\varepsilon_0^2}\frac{1}{n^2} \tag{53a}$$

which gives an exciton Rydberg constant

$$R_e = \frac{2\pi^2 e^4 \mu_0}{h^3 c\varepsilon_0^2} = 109\,737\,\frac{\mu_0}{m_0\varepsilon_0^2}\,\text{cm}^{-1}. \tag{53b}$$

The extreme example of anisotropy occurs when the electron, hole effective masses in the c direction are infinite; this corresponds, (51), to $\gamma = \frac{3}{2}$. Taking $\mu_\parallel$ to be infinite (50a) becomes

$$\left\{-\frac{\hbar^2}{2\mu_\perp}\left(\frac{\partial^2}{\partial x^2}+\frac{\partial^2}{\partial y^2}\right)-\frac{e^2}{\varepsilon_0}(x^2+y^2+z'^2)^{-(1/2)}\right\}\psi = E\psi. \tag{54a}$$

Solutions of (54a) are of the form $\psi(x, y)\,\delta(z-z_0)$ where z_0 is a quantum number, this means that the electron-hole separation is constant in the z direction. The energies of such bound pair states are given by [21, 22].

$$E_n = E_g - \frac{R}{(n+\tfrac{1}{2})^2}, \qquad n = 0, 1, 2, \ldots, \tag{54b}$$

where R is the exciton Rydberg energy and the line oscillator strengths [23] vary as $(n+\tfrac{1}{2})^{-3}$. The two dimensional model of bound, and unbound, electron-hole states has been applied to layer structures but it has been suggested [24] that a two-dimensional representation is an inadequate description of the states of a physical system.

Because of the large radii of delocalized excitons, (52c), the situation can arise in which the diameter of the exciton is comparable with the crystal dimensions. This can occur in evaporated films [25], where the crystallite size is very small, and in thin cleaved single crystals. The case of exciton confinement in a thin crystal has been treated [20] by regarding the crystal as an isotropic dielectric medium bounded by planes $z = \pm L$ and unlimited in the x and y directions. The chosen boundary conditions were that for either electron coordinate $\mathbf{r}_e$ or hole coordinate $\mathbf{r}_h$ outside certain effective crystal boundaries $\psi(\mathbf{r}_e, \mathbf{r}_h) = 0$. If the exciton centre of mass is constrained to lie at the crystal centre then for $r^{(n)}/L \ll 1$ the predicted shift of energy levels as a result of exciton confinement is as shown in Figure 8 which was calculated using the exciton parameters appropriate to

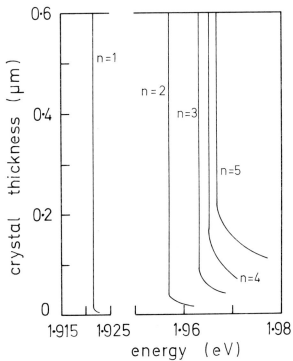

Fig. 8. Predicted shift of exciton energy levels with crystal thickness [20]. Numerical values refer to MoS$_2$[2].

MoS$_2$ [2]. For a given crystal thickness the energy shift ΔE increases with n (i.e. exciton diameter) so that the n^{-2} energy dependence in thick crystals, (44, 52, 53), does not apply to thin crystals. The predicted variation of ΔE with crystal thickness, Figure 8, is more rapid than the observed variation [26] while if the exciton centre of mass is allowed complete freedom within the crystal then calculation gives an L^{-1} dependence of ΔE which is too slowly varying. When the crystal thickness < exciton Bohr diameter then the problem simplifies to give [20, 27] $E \propto L^{-2}$ which is the limiting L dependence of ΔE.

3.2. Delocalized (M_0) excitons: optical selection rules [28]

The expression (44) for the exciton binding energy was derived assuming parabolic valence and conduction bands having extrema at $\mathbf{k} = 0$. Allowed direct transitions across the energy gap E_g between such valence and conduction bands gives rise to an M_0 interband absorption edge. Figure 1.

The probability per unit time that a transition occurs between the ground state ψ_0 of the crystal (volume V) and a discrete exciton state $\psi_{\nu\mathbf{K}}$ is calculated, c.f. (1), by assuming that the radiation covers a spread of frequencies. Thus if the intensity in the small frequency range $\Delta\omega$ is $I(\omega)\Delta\omega$ then the vector potential $\mathbf{A}_0$ characterizing the frequency range $\Delta\omega$ is given by

$$|\mathbf{A}_0|^2 = \frac{2\pi c}{n\omega^2} I(\omega)\Delta\omega, \tag{55}$$

where n is the refractive index. The transition probability per unit time per unit crystal volume is then

$$\frac{w}{V} = \frac{4\pi^2 e^2}{m^2 cn V \omega_{\nu,0}} I(\nu, 0) |H''_{\nu,0}|^2, \tag{56a}$$

where

$$H''_{\nu,0} = \int \psi^*_{\nu,\mathbf{K}} \alpha \sum_j e^{i\mathbf{q}\cdot\mathbf{r}_j} \nabla_j \psi_0 \, d\tau_1 \cdots d\tau_{2N} \tag{56b}$$

and the sum on j runs over the valence electrons. In the exciton representation $\psi_{\nu,\mathbf{K}}$ is written as the linear combination [13]

$$\psi_{\nu,\mathbf{K}} = \sum_\beta U_{\nu\mathbf{K}}(\beta) \Phi_{vc}(\mathbf{K}, \beta) \tag{57a}$$

where the U are coefficients and

$$\Phi_{vc}(\mathbf{K}, \beta) = N^{-(1/2)} \sum_\mathbf{R} e^{i\mathbf{K}\cdot\mathbf{R}} \Phi_{vc}(\mathbf{R}, \mathbf{R}+\beta) \tag{57b}$$

the latter representation conforming to the picture of a hole at one lattice site ($\mathbf{R}$) and an electron at another, distant β, the pair moving with total momentum $\mathbf{K}$.

Substituting (57) in (56) and expressing ψ_0 and $\Phi(\mathbf{R}, \mathbf{R}+\beta)$ in terms of localized functions gives

$$H''_{\nu,0} = \left[N^{-(1/2)} \sum_\mathbf{R} e^{i(\mathbf{q}-\mathbf{K})\cdot\mathbf{R}}\right] \alpha \int \psi^*_{c,\nu}(\mathbf{r}) e^{i\mathbf{q}\cdot\mathbf{r}} \nabla a_{v,0}(\mathbf{r}) \, d\tau, \tag{58a}$$

where

$$\psi^*_{c,\nu} = \sum_\beta U_{\nu \mathbf{K}}(\beta) a_{c\beta}(\mathbf{r}) \tag{58b}$$

and $a_{\nu,0}$ is the initial state of an electron which makes a transition to state $\psi_{c,\nu}$. Equation (58a) is non-zero only when $\mathbf{K} = \mathbf{q}$. Since for visible light, $\mathbf{q}$ is near the centre of the Brillouin zone this condition reduces to

$$\mathbf{K} \simeq 0 \tag{59}$$

so that transitions occur only to the bottom ($\mathbf{K} = 0$) of each exciton band giving, (45), an exciton line spectrum.

For dipole allowed transitions only the first term, viz unity, in the expansion of $\exp(i \cdot \mathbf{q} \cdot \mathbf{r})$ need be considered. Equation (56a) can then be used to obtain an expression for the integrated absorption, the magnitude of which is mainly determined by the square of a term the z component of which is

$$z_{\nu 0} = \int \psi^*_{c\nu} z \, a_{\nu 0}(\mathbf{r}) \, d\mathbf{r} = \sum_\beta U_{\nu \mathbf{K}}(\beta) \int a^*_{c\beta}(\mathbf{r}) \, z a_{\nu 0}(\mathbf{r}) \, d\mathbf{r} \tag{60}$$

Extracting the $\beta = 0$ term (in which the electron and hole are on the same lattice site) from the summation (60) becomes

$$z_{\nu 0} = U^*_{\nu \mathbf{K}}(0) \int a^*_{c0}(\mathbf{r}) z a_{\nu 0}(\mathbf{r}) + \sum_{\beta \neq 0} U^*_{\nu \mathbf{K}}(\beta) \int a^*_{c\beta}(\mathbf{r}) z a_{\nu 0}(\mathbf{r}) \, d\mathbf{r}. \tag{61}$$

Two cases arise

(a) *First class or 'allowed' transitions* occur when

$$z_{cv} = \int a^*_{c0}(\mathbf{r}) z a_{\nu 0}(\mathbf{r}) \, d\mathbf{r} \neq 0. \tag{62}$$

In this case the $\beta = 0$ term in (61) dominates the summation and the remaining $\beta \neq 0$ terms can be considered two centre corrections. The integrated absorption, i.e. transition oscillator strength per electron is then given by

$$f = \frac{2m}{\hbar^2} (E_{\nu \mathbf{K}} - E_0) |U_{\nu \mathbf{K}}(0)|^2 |z_{cv}|^2, \tag{63a}$$

where E_0 is the first order ground state energy.

It can be shown that $U_{\nu \mathbf{K}}(0)$ is proportional to the s-like envelope functions for hydrogenic excitons so that

$$f = \frac{2m}{\hbar^2} (E_{\nu \mathbf{K}} - E_0) V_c \left| \frac{z_{cv}}{\pi r_e^3} \right|^2 \cdot \frac{1}{n^3}, \tag{63b}$$

where V_c, the unit cell volume is a normalization factor, $r_e = \epsilon \hbar^2 / \mu e^2$ is the radius of the $n = 1$ exciton orbit and quantum numbers $n = 1, 2, 3$ etc. for the s-state excitons. The wavenumbers at which the exciton lines occur are given by, (45)

$$\nu_n = \nu_g - \frac{R_e}{n^2}, \quad n = 1, 2, 3, \ldots, \tag{64}$$

where R_e is given by (53b) in anisotropic crystals.

Because of the rapid fall off in intensity the high quantum number $(n>3)$ exciton lines are difficult to detect experimentally. At large values of n the lines are so closely spaced, (64), that they resemble a continuum; since the line density increases as n^3 and the line intensity as n^{-3} the absorption coefficient of this *overlap continuum* is independent of frequency. The overlap continuum extends to $\nu = \nu_g(n=\infty)$. At $\nu > \nu_g$ the true continuum corresponding to ionized pair states occurs. For photon energies only slightly in excess of the band gap energy E_g

$$\Delta E = \hbar\omega - E_g \tag{65a}$$

the absorption coefficient in the continuum is given by

$$\alpha = \frac{2\pi\omega\varepsilon}{r_e^2 cn}|z_{cv}|^2 \frac{\exp \pi x}{\sinh \pi x}, \tag{65b}$$

where $x^2 = hcR_e/\Delta E$. When $\Delta E \gg hcR_e$ (65b) reduces to $\alpha \propto (\Delta E)^{1/2}$ which is the same as the absorption edge for direct interband transitions at an M_0 c.p (21).

(b) *Second class or 'forbidden' transitions* occur when a_{c0} and a_{v0} have the same parity so that z_{cv}, (62), is zero. In this case, (61),

$$z_{\nu 0} = \sum_{\beta=0} U^*_{\nu\mathbf{K}}(\beta) \int a^*_{c\beta}(\mathbf{r}) z a_{v0}(\mathbf{r})\, d\mathbf{r}. \tag{66}$$

If $+\boldsymbol{\beta}_1$ and $-\boldsymbol{\beta}_1$ are the positions along the z axes (strictly of a cubic crystal but the results can be generalized) of the two neighbours nearest the origin then [29]

$$z_{\nu 0} \simeq [U_{\nu\mathbf{K}}(\boldsymbol{\beta}_1) - U_{\nu\mathbf{K}}(-\boldsymbol{\beta}_1)] z_{cv}(\boldsymbol{\beta}_1), \tag{67}$$

where $z_{cv}(\boldsymbol{\beta}_1)$ is the two centre integral of (66). If $\boldsymbol{\beta}_1$ is small enough to be written as a differential then writing

$$[U_{\nu\mathbf{K}}(\boldsymbol{\beta}_1) - U_{\nu\mathbf{K}}(-\boldsymbol{\beta}_1)]/2\boldsymbol{\beta}_1 = \partial U_{\nu\mathbf{K}}(0)/\partial \beta_z \tag{68}$$

(67) leads to the following expression for the transition oscillator strength viz.

$$f = \frac{2m}{\hbar^2}(E_{\nu\mathbf{K}} - E_0)|2\boldsymbol{\beta}_1 z_{cv}(\boldsymbol{\beta}_1)|^2 \left|\frac{\partial}{\partial \beta} U_{\nu\mathbf{K}}(0)\right|^2 \tag{69a}$$

which is non-vanishing only for p-like envelope functions. Substituting the appropriate normalized hydrogenic function (69a) becomes

$$f = \frac{2m}{\hbar^2}(E_{\nu\mathbf{K}} - E_0)|2\boldsymbol{\beta}_1 z_{cv}(\boldsymbol{\beta}_1)|^2 \frac{V_c}{3\pi r_e^5}\left(\frac{n^2-1}{n^5}\right). \tag{69b}$$

Since (69b) is non-zero only for p-like envelope functions the quantum number $n = 2, 3, 4$ etc. The wavenumber positions of the second class exciton lines are given by

$$\nu_n = \nu_g - \frac{R_e}{n^2}, \quad n = 2, 3, 4, \ldots \tag{70}$$

Comparing (69) and (63b) it is evident the 'forbidden' transitions are weaker than 'allowed' transitions by at least a factor of $(\beta_1/r_e)^2$. In the true continuum $(\hbar\omega = E_g + \Delta E)$ the absorption coefficient is given by

$$\alpha = \frac{2\pi\omega\varepsilon}{3r_e^5 cn} |2\beta_1 z_{cv}(\beta_1)|^2 \frac{e^{\pi x}(1+x^2)}{2\sinh \pi x} \tag{71}$$

which reduces to a value appropriate to the overlap continuum when $\Delta E \to 0$ and to the free particle form $\alpha \propto (\Delta E)^{3/2}$ when $\Delta E \gg hcR_e$.

Dipole forbidden transitions

If a dipole transition is forbidden by symmetry then the integral in (58a) is approximately given by the second term in the expansion of $\exp(i\mathbf{q}\cdot\mathbf{r})$. Viz.

$$\alpha \int \psi_{c,v}^*(\mathbf{r}) i\mathbf{q}\cdot\mathbf{r} \nabla a_{v,0}(\mathbf{r}) \, d\tau. \tag{72}$$

This is small so that the associated exciton line is very weak; experimentally the line may be undetectable.

3.3. Delocalized hyperbolic (M_1, M_2) excitons

In principle exciton states can be associated with every type of critical point since the condition for a critical point, (15a) is identical with the requirement that, for an exciton, the electron and hole have the same group velocity. The excitons predicted [30] at M_1 and M_2 critical points are described as hyperbolic or saddle point excitons.

One approach to the problem of calculating how exciton effects modify the form of the absorption at M_1, M_2 critical points is to assume that the effective mass equation viz:

$$\left[\frac{-\hbar^2}{2\mu_1'}\frac{\partial^2}{\partial x^2} - \frac{\hbar^2}{2\mu_2'}\frac{\partial^2}{\partial y^2} - \frac{\hbar^2}{2\mu_3'}\frac{\partial^2}{\partial z^2} + V(r)\right]\psi = E\psi \tag{73}$$

still applies; here the reduced electron-hole mass is redefined by

$$\frac{1}{\mu'} = \frac{1}{m_e^*} - \frac{1}{m_h^*}, \tag{74}$$

where m^* can take positive or negative values.

At M_1 (M_2) critical points one (two) components of μ' are negative and no bound solutions of (74) exist when $V(r)$ is represented by a pure Coulomb potential or one of several separable approximations to the Coulomb potential [31]. Some indication of the behaviour of ε_2 near an M_1 critical point can be obtained by considering the situation when the negative reduced mass, μ_3 say, becomes infinite. In this case the cylindrically symmetric form of (73) becomes

$$\left[-\frac{\hbar^2}{2\mu_\perp'}\left(\frac{\partial^2}{\partial x^2} + \frac{\partial^2}{\partial y^2}\right) - \frac{e^2}{\varepsilon(x^2+y^2+z^2)^{1/2}}\right]\psi = E\psi. \tag{75}$$

The eigenfunctions have the form

$$\psi = \phi_n(x, y : \xi) \, \delta(z - \xi), \tag{76}$$

where ϕ is a solution of (75) if the variable z is treated as a parameter with fixed ξ on which the functions ϕ_n and corresponding eigenvalues $E_n(\xi)$ depend. Substituting $z = 0$ (75) becomes the Schrödinger equation of the two dimensional hydrogen atom which has bound states at energies, below E_g, of [32]

$$E_n(0) = -\frac{2\mu'_\perp e^4}{\hbar^2 \varepsilon^2 (2n-1)^2}; \quad \text{for which} \quad |\phi_n(0)|^2 = \frac{16(\mu'_\perp)^3 e^6}{\hbar^6 \pi \varepsilon^3 (2n-1)^3}. \tag{77}$$

In the continuum above E_g

$$|\phi_E(0)|^2 = \frac{\exp(-\gamma')}{\cosh \gamma'} \quad \text{where} \quad \gamma' = \frac{-\pi e^2 \sqrt{\mu'_\perp}}{\hbar \varepsilon [2(E - E_g)]^{1/2}}. \tag{78}$$

Equations (77, 78) give the spectral variation of ε_2 shown in Figure 9. The effect of (negative) $\mu'_\parallel$ being large, instead of infinite, is to broaden the δ function exciton peak [33].

At an M_2 critical point two components of μ' are negative. An asymptotic treatment of this problem [33] in which the positive component of $\mu' \to \infty$ predicts (a) the absence of bound states (b) that the $e-h$ interaction has the effect of reducing the interband absorption to a minimum at $E = E_g$.

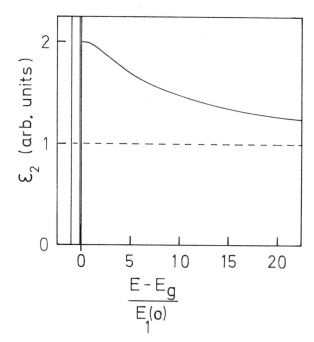

Fig. 9. The spectral variation of ε_2 near a two dimensional M_0 critical point when electron-hole interaction is included. A similar variation is expected near a three dimensional M_1 critical point when the magnitude of the negative reduced mass is much larger than that of the positive masses [33].

3.4. EXCITON EFFECTS AT AN M_3 CRITICAL POINT

At an M_3 critical point all three components of μ' are negative and the solutions of (73) are similar to those obtained for a repulsive Coulomb field and $\mu' > 0$. There are no bound states and the spectral variation of ε_2 near the critical point is given by [33]

$$e_2 \propto \beta \frac{\exp \beta}{\sinh \beta}, \tag{79a}$$

where

$$\beta = \frac{-\pi e^2 \sqrt{\mu'}}{\varepsilon \hbar [2(E_g - E)]^{1/2}}. \tag{79b}$$

Comparing (79) with the expression for ε_2 due to allowed interband transitions at an M_3 critical point, Figure 1, it is seen that the $e-h$ interaction weakens the M_3 singularity to such an extent that it will be difficult to detect.

3.5. INTERMEDIATE EXCITONS

In the intermediate, Type III, class of crystal, Section 3, the short range interaction potential U is non-zero only when the Wannier electron and hole are in the same unit cell. On this scheme the electron-hole interaction is given by the matrix elements $\langle cv\mathbf{k}|U|c'v'\mathbf{k'}\rangle = N^{-1}g$ where $g<0$ (attractive forces) is the coupling constant of the interaction and N the number of unit cells in the crystal, volume V. For such an interaction potential (g small) the dielectric constant is given by [33]

$$\varepsilon_2(\omega) \simeq [1 - 2g \, \text{Re} \, F(\omega)] \varepsilon_2^0(\omega), \tag{80a}$$

where $\varepsilon_2^0(\omega)$ corresponds to the band approximation ($g=0$) and

$$\text{Re} \, F(\omega) = -\frac{V}{N} \int_{-\infty}^{\infty} \frac{d\xi}{\omega - \xi} J_{cv}(\xi), \tag{80b}$$

where $J_{cv}(\omega)$ is the joint density of states. In order to determine $\varepsilon_2(\omega)$ from $\varepsilon_2^0(\omega)$ it is only necessary to know $J_{cv}(\omega)$. The form of $\varepsilon_2(\omega)$ near critical points M_t for particular g values is shown in Figure 10 which also demonstrates the effect of increasing electron-hole interaction (increasing g) [34]. At three dimensional critical points M_t the complex dielectric constant (for $g=0$) is given by (27) viz.

$$\hat{\varepsilon} \propto b(\omega - \omega_0)^{1/2} + \text{constant}, \tag{81}$$

where $b = i^{t+1}$. The form of ε_2 (for $g = 0$) at each type of critical point is shown in Figure 10 against its corresponding b value in the complex plane, cf. Figure 1. Increasing $|g|$, ($g<0$) has the effect of rotating b anticlockwise so that $\varepsilon_2(\omega)$ at an M_t critical point takes on more and more the form appropriate to an $M_{t+1}(g=0)$ critical point. The form of $\varepsilon_2(\omega)$, (80), for a g value corresponding to a rotation of

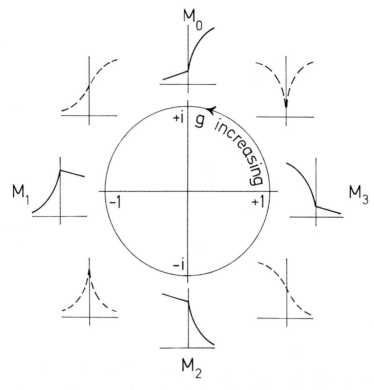

Fig. 10. The form of $\varepsilon_2^0(\omega)$ near three-dimensional Van-Hove singularities [34]. A small, short range exciton interaction has the effect of mixing the M_t and M_{t+1} singularities as shown by the dashed examples.

ca. 45° is shown dotted in Figure 10; one obvious feature is the way in which the interaction sharpens the absorption structure (to give a peak) near an M_1 edge. At large values of g, such that $|gF(\omega)| \geq 1$ the approximate equation, (80), is no longer valid and the calculated [35] $\varepsilon_2(\omega)$ curves are very different to those in Figure 10.

Short range (contact) models of the interaction have the property that interactions strong enough to bind a parabolic exciton in the energy gap are too strong to produce a hyperbolic exciton. Such models can, in fact, give rise to only one resonance since, in the Wannier representation, the $e-h$ interaction matrix has only one non-vanishing element [36].

3.6. INDIRECT EXCITON TRANSITIONS

If lattice vibrations are no longer excluded from the crystal model then optical transitions to exciton states $\mathbf{K} \neq 0$ can occur by the simultaneous emission or absorption (at $T>0$) of a phonon whose wave vector $\mathbf{k}_{ph} \approx \mathbf{K}$. Since a large range of $\mathbf{k}_{ph}$ is available transitions can occur to any point of the exciton band. Thus indirect transitions result in a continuous spectrum rather than the line spectrum characteristic of direct transitions, 3.2.

Consider a single crystal having an isotropic valence (conduction) band with maximum (minimum) at $\mathbf{k}=0$ ($\mathbf{k}=\mathbf{k}_c$). Absorption due to indirect transitions into the associated bound exciton states will be proportional to the density of states which, since (for given n) only a single parabolic band of final states exists, is given by

$$\int d^3 K \, \delta\left\{ E'_g + \frac{\hbar^2 K^2}{2M} - \frac{hc}{n^2} R_e(\mathbf{k}_c) \pm \hbar\omega_l(\mathbf{k}_{ph}) - \hbar\omega \right\}$$

and to the probability of forming an exciton (which is proportional to n^{-3} for 'allowed' transitions, (63b), and $(n^2-1)/n^5$ for 'forbidden' transitions). Thus for indirect 'allowed' exciton transitions the absorption coefficient (for $\hbar\omega \simeq E'_g$, the indirect band gap) is given by [28]

$$\alpha \propto \frac{1}{\hbar\omega} \sum_{n=1}^{\infty} \frac{1}{n^3} \left\{ \hbar\omega - E'_g + \frac{hc}{n^2} R_e(\mathbf{k}_c) \pm \hbar\omega_l(\mathbf{k}_{ph}) \right\}^{1/2}. \tag{82}$$

The indirect absorption spectrum therefore consists of a series of steps, one for each exciton state n and two for each phonon branch l corresponding to either the creation or destruction of a phonon. Because of the n^{-3} dependence the first, $n=1$, step will normally be dominant. The absorption coefficient due to indirect transitions into exciton bound states is very small and can usually only be detected at the absorption threshold, although even here it can be masked by free carrier absorption.

The absorption coefficient associated with 'allowed' indirect transitions into unbound exciton states is, for $\hbar\omega \simeq E'_g$, given by [28]

$$\alpha\omega \propto [\hbar\omega - E'_g \pm \hbar\omega_l(\mathbf{k}_{ph})]^{3/2} \tag{83}$$

so that the absorption rises as the three-halves power of the energy above threshold compared with the (energy)^2 dependence of indirect interband transitions, (35).

3.7. EXCITON LINE BROADENING

Lattice vibrations broaden the line (δ function) exciton absorption spectrum of the rigid lattice. In the case of an isolated exciton energy band it is evident that phonon assisted optical transitions can occur to points $\mathbf{K} \neq 0$ with the result that the original δ function absorption line is asymmetrically broadened on its high energy side.

When the complete exciton band structure of the crystal is considered then interband transitions (involving intermediate states in different (n) bands) can also occur. Due to the electrical neutrality of the exciton the optical mode of vibration is not important in the intraband scattering. In the interband scattering however, in which the quantum number of the exciton changes, the contribution of the optic mode may be as large as that of the acoustic mode [37].

The way in which an exciton absorption line is modified as a result of intraband transitions alone has been calculated [38] for the limiting cases of (1) weak and (2) strong exciton-phonon interactions.

(1) For weak interactions then provided the exciton energy band extremum does not occur at $K = 0$ the absorption coefficient due to the nth exciton (assuming the refractive index is constant) is given by

$$\alpha_n(\omega) \propto \frac{\hbar \gamma_n/2}{[\hbar\omega - (E_n + \Delta_n)]^2 + (\hbar \gamma_n/2)^2}, \tag{84}$$

where Δ_n is an energy shift. Equation (84) shows that the exciton absorption has a Lorentzian shape with a peak at $E_n + \Delta_n$ and a half-width $H = \hbar \gamma_n$ which represents the broadening. Except at low temperatures $H \propto T$.

Equation (84) does not apply when the exciton band has a maximum or minimum at $\mathbf{K} = 0$; in such cases multiphonon processes are important. If the exciton energy band has a minimum at $\mathbf{K} = 0$ then the associated absorption band is very asymmetric [39] having an extended tail on the the high energy side of maximum but being reduced to zero at an energy ca. $\frac{1}{4}$ half width below peak energy.

(2) For strong electron-phonon interaction (or when T is very large) the exciton absorption band shape due to intraband transitions is (provided the band extrema are at $K \neq 0$) given by

$$\alpha_n(\omega) \propto \frac{1}{D} \exp\left[-\frac{(\hbar\omega - E_n)^2}{2D^2}\right] \tag{85}$$

which is of Gaussian shape with a peak at E_n and a half-width $H = 2^{3/2}(\ln 2)D$ where $H \propto \sqrt{T}$.

When interband effects are taken into account the line shape in the weak coupling case, (84), is modified into an asymmetric Lorentzian [38, 40]

$$\alpha_n(\omega) \propto \frac{(\hbar \gamma_n/2) + 2A_n[\hbar\omega - (E_n + \Delta_n)]}{[\hbar\omega - (E_n + \Delta_n)]^2 + (\hbar \gamma_n/2)^2}, \tag{86}$$

where the ω dependence of the asymmetry A and damping γ are neglected. The half-width $H \propto T$.

As a result of the broadening of the excition absorption lines due to intra and interband transitions considerable overlap occurs between the absorption bands of even low quantum number ($n = 1, 2$ etc.) excitons.

In order to reduce the overlap to a minimum, and thereby detect the component bands, it is necessary to cool the crystal to a low temperature.

4. Free Carrier Absorption

In semiconductors the free carriers present in the crystal give rise to absorption and dispersion, effects which become significant to the low energy side of the absorption threshold. Theories of varying sophistication have been developed to describe the free carrier absorption as follows.

4.1. CLASSICAL MODEL

The Drude-Zener equations [41, 42, 43] can be used to describe the absorption in which case

$$\alpha = \frac{4\pi Ne^2}{nmc} \frac{1/\tau}{\omega^2 + (1/\tau)^2}, \tag{87a}$$

where τ is an electron–lattice ion collision time. For $\omega\tau \gg 1$ (87a) becomes

$$\alpha = \frac{4\pi Ne^2}{nmc\tau^2} \frac{1}{\omega^2}. \tag{87b}$$

Equations (87) are not very satisfactory since they assume that τ is the same for all electron energies.

4.2. ELECTRON ENERGY BAND MODEL

On this model free carrier absorption corresponds to excitation of (holes) electrons within the same partly filled band i.e. intraband transitions. Conservation of crystal momentum requires that intraband transitions are accompanied by scattering of phonons. In the case of isotropic scattering this quantum theory of free carrier absorption gives, for $\hbar\omega < kT$, results [44] which are within 10% of those given by the classical equation (87).

For spherical energy surfaces the energy E_f of the final state achieved by an intraband transition, via intermediate state E_i, from an initial state E_0 is

$$E_f = E_0 + \frac{\hbar^2}{2m^*}(k_f^2 - k_0^2) \pm \hbar\omega_{ph} - \hbar\omega. \tag{88}$$

The associated absorption coefficient is given by [45]

$$\alpha = \frac{4\pi N}{cn} \frac{4e^3 \pi^{1/2}}{9m^2\omega^2} \left(\frac{\hbar\omega}{kT}\right)^{1/2} \frac{1}{\mu_a}, \tag{89}$$

where μ_a is the mobility corresponding to acoustic mode scattering. Equation (89) predicts an $\omega^{-(3/2)}$ dependence rather than the ω^{-2} dependence of (87).

A more extensive treatment of intraband absorption has been given [46] for ellipsoidal energy surfaces which includes the effects of intravalley and direct and indirect intervalley transitions. The general expression for the intraband absorption resulting from these various phonon processes is [46]

$$\alpha = \frac{4\pi}{nc}(\sigma_1 + \sigma_2 + \sigma_3), \tag{90a}$$

where σ_1 and σ_2 are due to processes involving, respectively, transverse and longitudinal long wavelength acoustic phonons and σ_3 contains all energetic mode contributions. Writing $\nu = \omega/2\pi$ and $x = h\nu/kT$

$$\sigma_1(\nu) = \Gamma_1(kT)^{-(1/2)}x^{-1}\sinh(x)K_2(x), \tag{90b}$$

$$\sigma_2(\nu)/\sigma_1(\nu) = \Gamma_2/\Gamma_1, \tag{90c}$$

$$\sigma_3(\nu)/\sigma_1(\nu) = G, \tag{90d}$$

where G is effectively independent of ν and T at sufficiently large ν, T. The constants Γ_1, Γ_2 (for a given crystal) can be expressed in terms of deformation potentials [47] and $K_2(x)$ is a modified Bessel function [48].

5. The Effect of an Applied Electric Field on the Crystal Dielectric Function

In many cases the structure present in the conventional transmission (reflection) spectrum of a crystal is not sufficiently well defined as to allow positive identification of the associated critical point transition. For this reason measurements are made of the effect of an applied field on the spectrum, this often allows the symmetry of the transition to be determined.

The way in which an electric field modifies the crystal dielectric function in the vicinity of a critical point has been determined in the effective mass approximation (EMA) [49, 50] and also by a more exact one-electron treatment which takes into account the lattice periodicity; this latter treatment (using the equivalent [51] Houston [52–56] and crystal momentum representation (CMR) methods [57–60] reduces to the results obtained in the EMA method and also predicts the existence, so far unconfirmed, of fine structure known as Stark steps [57].

5.1. Effective mass approximation (EMA)

The effective mass equation for an electron-hole pair in a uniform electric field $\boldsymbol{\xi}$ is

$$\left\{\frac{\hbar^2}{2}\left[\frac{1}{\mu_x}\frac{\partial^2}{\partial x^2} + \frac{1}{\mu_y}\frac{\partial^2}{\partial y^2} + \frac{1}{\mu_z}\frac{\partial^2}{\partial z^2}\right] + e\boldsymbol{\xi}\cdot\mathbf{r} + W\right\}\varphi(\mathbf{r}) = 0, \tag{91}$$

where, in order to obtain a closed form analytic solution, the Coulomb interaction has been neglected by considering only crystals with weak exciton binding. The solution $\varphi(\mathbf{r})$ of (91) can be written as the product of the three functions $\varphi(r_i)$ each satisfying the equation

$$\left\{\frac{\hbar^2}{2\mu_i}\frac{\partial^2}{\partial r_i^2} + e\xi_i r_i + W_i\right\}\varphi(r_i) = 0 \tag{92}$$

in which case the eigenvalue of (91) is $W = W_1 + W_2 + W_3$. When the field component in the ith direction is zero then (92) has the plane wave solution $\varphi(r_i) \propto \exp(ik_i r_i)$.

When ξ_i is not zero then by means of the change in variables

$$\eta_i = (W_i + e\xi_i r_i)/\hbar\theta_i, \tag{93}$$

where $\theta_i^3 = e^2 \xi_i^2 / 2\hbar\mu_i$ (92) becomes

$$\frac{\partial^2}{\partial \eta^2} \varphi(\eta) = \pm \eta \varphi(\eta), \tag{94}$$

where the positive (negative) signs refer to negative (positive) μ. The solution of (94), regular at infinity, is [61]

$$\varphi(\eta_i) = C_i Ai(\pm \eta_i), \tag{95a}$$

where the Airy function is defined as

$$Ai(x) = \frac{1}{\pi} \int_0^\infty ds \cos\left(\frac{s^3}{3} + sx\right) = \frac{1}{2\pi} \int_{-\infty}^\infty ds \exp\left(i\frac{s^3}{3} + isx\right). \tag{95b}$$

The normalization coefficient C_i is such that the wave function $\varphi(\mathbf{r})$ gives the delta function over energy for each coordinate i.e. [50]

$$C_i = (e|\xi_i|)^{1/2}/\hbar\theta. \tag{95c}$$

Substituting for $|\varphi(0)|^2$ in an expression for $\varepsilon_2(\omega)$ due to direct allowed transitions gives, c.f. (13),

$$\varepsilon_2(\omega, \xi) = \frac{4\pi^2 e^2}{m^2 \omega^2} |\boldsymbol{\alpha} \cdot \mathbf{p}_{cv}|^2 e^3 \frac{|\xi_x \xi_y \xi_z|}{\hbar^6 \theta_x^2 \theta_y^2 \theta_z^2}$$

$$\times \int_{-\infty}^\infty dW_x \, dW_y \, dW_z Ai^2\left(\frac{-W_x}{\hbar\theta_x}\right) Ai^2\left(\frac{-W_y}{\hbar\theta_y}\right) Ai^2\left(\frac{-W_z}{\hbar\theta_z}\right)$$

$$\times \delta(E_g + W_x + W_y + W_z - \hbar\omega), \tag{96}$$

where $Ai^2(x) = [Ai(x)]^2$.

Equation (96) is a general expression for the behaviour of $\varepsilon_2(\omega, \xi)$ in the vicinity of the four different types of critical point (identified by the sign of θ_i, (93)). Examination of (96) for each of the critical points (c.p) shows that the expression for $\varepsilon_2(\omega, \xi)$ appropriate to an M_3 c.p is the same as that for an M_0 c.p except that $(E_g - \hbar\omega)$ is replaced by $-(E_g - \hbar\omega)$ i.e.

$$\varepsilon_2^{M_0}(\xi, -(E_g - \hbar\omega)) = \varepsilon_2^{M_3}(\xi, (E_g - \hbar\omega)). \tag{97a}$$

Similarly

$$\varepsilon_2^{M_1}(\xi, -(E_g - \hbar\omega)) = \varepsilon_2^{M_2}(\xi, (E_g - \hbar\omega)). \tag{97b}$$

On evaluating (96) it is found [50] that the results, expressed as the difference $\Delta\varepsilon_2(\omega, \xi) = \varepsilon_2(\omega, \xi) - \varepsilon_2(\omega, 0)$ can be given in terms of the two functions

$$F(x) = \pi[Ai'^2(x) - Ai^2(x)] - \sqrt{-x}\, D(-x), \tag{98a}$$

$$G(x) = \pi[Ai'(x)Bi'(x) - xAi(x)Bi(x)] + \sqrt{x}\, D(x), \tag{98b}$$

where $Ai'(x) = (d/dx)Ai(x)$, $Bi(x)$ is the solution of (94) which is irregular at infinity viz.

$$Bi(x) = \frac{1}{\pi}\int_0^\infty ds\left[\exp\left(\frac{s^3}{3}+xs\right) - \sin\left(\frac{s^3}{3}+xs\right)\right] \tag{98c}$$

and $D(x)$ is the step function equal to one for positive x and zero for negative x.

The complete expressions for $\Delta\varepsilon_2(\omega, \xi)$ at each of the different types of three dimensional c.p are listed in Table III and drawn in Figure 11 for specific values of the parameters. The six possible line shapes for $\Delta\varepsilon_2(\omega, \xi)$, Table III, are given by

TABLE III

The change in the dielectric function near a three dimensional c.p as a result of an applied electric field. The constants and functions employed are defined in the text

Critical point	Field direction (sign of $\hbar\theta_0$)	$\Delta\varepsilon_1(\omega, \xi)$	$\Delta\varepsilon_2(\omega, \xi)$
$M_0(\mu_x, \mu_y, \mu_z > 0)$	$\hbar\theta_0 > 0$	A.G(x)	A.F(x)
$M_1\begin{pmatrix}\mu_x, \mu_y > 0\\ \mu_z < 0\end{pmatrix}$	parallel $\hbar\theta_0 < 0$	A.G(−x)	−A.F(−x)
	transverse $\hbar\theta_0 > 0$	−A.F(x)	A.G(x)
$M_2\begin{pmatrix}\mu_x, \mu_y < 0\\ \mu_z > 0\end{pmatrix}$	parallel $\hbar\theta_0 > 0$	−A.G(x)	−A.F(x)
	transverse $\hbar\theta_0 < 0$	A.F(−x)	A.G(−x)
$M_3(\mu_x, \mu_y, \mu_z < 0)$	$\hbar\theta_0 < 0$	−A.G(−x)	A.F(−x)

$$\Delta\varepsilon_2(\omega, \xi) = \pm\begin{cases} AF(\pm x) \\ AG(\pm x) \end{cases}, \tag{99a}$$

where x is the dimensionless variable $(E_g - \hbar\omega)/\hbar\theta_0$ and

$$\theta_0^3 = \theta_x^3 + \theta_y^3 + \theta_z^3 = \frac{e^2|\xi|^2}{2\hbar\mu_\xi}, \tag{99b}$$

$$A = \frac{2e^2}{\hbar m^2\omega^2}|\boldsymbol{\alpha}\cdot\mathbf{p}_{cv}(0)|^2\left[\frac{8}{\hbar^3}|\mu_x\mu_y\mu_z|\right]^{1/2}|\theta_0|^{1/2}, \tag{99c}$$

where μ_ξ is defined with respect to the principal axes of the effective mass tensor as

$$\frac{1}{\mu_\xi} = \frac{1}{|\xi|^2}\left\{\frac{\xi_x^2}{\mu_x} + \frac{\xi_y^2}{\mu_y} + \frac{\xi_z^2}{\mu_z}\right\} \tag{99d}$$

and is either positive or negative depending upon the c.p type and the direction of ξ. The dipole matrix element between the (parabolic) conduction and valence bands is assumed constant over the range of interest.

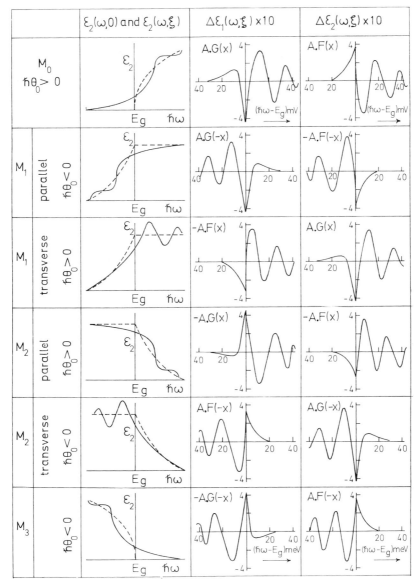

Fig. 11. The dielectric functions $\varepsilon_2(\omega, 0)$, dashed line, and $\varepsilon_2(\omega, \xi)$, full line, in the vicinity of three dimensional critical points. Columns three and four show $\Delta\varepsilon_1(\omega, \xi)$ and $\Delta\varepsilon_2(\omega, \xi)$, calculated [50] for $\hbar\theta = 10 \text{ meV}$, $E_g = 0.8 \text{ eV}$ and $A = |\theta_0|^{1/2}/\omega^2$. [456]

Column three of Figure 11 shows the $\Delta\varepsilon_1(\omega, \xi)$ traces derived from the associated $\Delta\varepsilon_2(\omega, \xi)$ traces, column four, by Kramers-Krönig transformation [50]. In the case of the M_1 and M_2 c.p's the form of the $\Delta\varepsilon_2(\omega, \xi)$ traces is strongly dependent on field direction, the oscillations in $\Delta\varepsilon_2(\omega, \xi)$ and $\Delta\varepsilon_1(\omega, \xi)$ can occur either above or below the c.p energy. This directional dependence is weaker in the case of $M_0(M_3)$ edges where μ_ξ is always positive (negative). Experimentally the oscillations in $\Delta\varepsilon_2(\omega, \xi)$ are dampened by broadening effects and the singularity at the c.p energy is usually the only structure detected. Just as the absorption threshold is the most distinctive feature of the $\varepsilon_2(\omega, 0)$ spectrum so electric field effects are most obvious here. In the case of an M_0 threshold the applied electric field produces an exponential tail in the absorption coefficient below threshold, Figure 11, which is readily observed as a shift in the absorption threshold to lower energies, the Franz-Keldysh effect. [53, 54, 49, 62.]

Expressions for $\varepsilon_2(\omega, \xi)$ near the two-dimensional c.p's which are thought to occur in layer-type crystals have been obtained [63] by a one-electron treatment (Houston method) which, as mentioned earlier, can also be employed to derive Figure 11. Let the energy expansion around the two-dimensional c.p be, c.f (16),

$$E_c - E_v = E_0 + a_x k_x^2 + a_y k_y^2, \tag{100}$$

where the signs of the coefficients 'a' determine the c.p type. An electric field applied perpendicular to a crystal layer has no effect on the optical constants in the present approximation, consequently it is assumed that the applied field is in the plane of a layer in a direction making an angle θ with the k_x axis. Defining

$$a_\| = a_x \cos^2 \theta + a_y \sin^2 \theta \tag{101}$$

it is found [63] that in the vicinity of a two-dimensional c.p

$$\varepsilon_2(\omega, \xi) = \frac{8\pi^2 e^2 \hbar^2}{m^2 \omega^2} |\boldsymbol{\alpha} \cdot \mathbf{p}_{cv}|^2 J_{cv}(\omega, \xi), \tag{102a}$$

where

$$J_{cv}(\omega, \xi) = \frac{1}{2\pi d} a_\|^{1/6} \frac{(e\xi)^{1/3}}{\sqrt{a_x a_y}} \int_p^q dt \frac{Ai^2(t)}{[|a_\||^{1/3} (e\xi)^{2/3} t - E_0 + \hbar\omega]} \tag{102b}$$

and d is the distance between layers.

For $M_0(2D)$ the limits of integration in (102b) are

$$p = (E_0 - \hbar\omega)/a_\|^{1/3}(e\xi)^{2/3} \quad \text{and} \quad q = (R^2 + E_0 - \hbar\omega)/a_\|^{1/3}(e\xi)^{2/3}.$$

For $M_1(2D)$

$$p = -(R^2 + E_0 - \hbar\omega)/|a_\||^{1/3} (e\xi)^{2/3}$$
$$q = -(E_0 - \hbar\omega)/|a_\||^{1/3} (e\xi)^{2/3} \quad \text{and} \quad |a_\||^{1/3}$$

in the denominator of (102b) is replaced by $-|a_\||^{1/3}$.

For $M_2(2D)$ the result can be obtained from that at $M_0(2D)$ by changing the sign of $E_0 - \hbar\omega$.

The form of $J_{cv}(\omega, \xi)$ near each two dimensional c.p is shown in Figure 12.

At an M_1 c.p the electro-optic effect depends upon the orientation of ξ with respect to the direction of negative effective mass, Figures 11 and 12. It follows

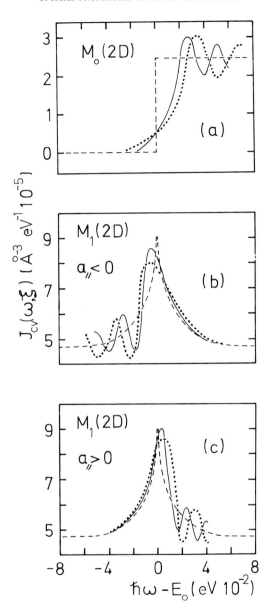

Fig. 12. [63] (a) $J_{cv}(\omega, \xi)$ at $M_0(2D)$. Values of the effective masses correspond to $a_x = 1.05 \times 10^{-3}$ eV^2 s^2 g^{-1}, $a_y = 0.26 \times 10^{-3}$ eV^2 s^2 g^{-1}. Full line is for $a_{\parallel}^{1/3}(e\xi)^{2/3} = 1.24 \times 10^{-2}$ eV; dotted line for $a_{\parallel}^{1/3}(e\xi)^{2/3} = 1.67 \times 10^{-2}$ eV; dashed line for $\xi = 0$. (b) $J_{cv}(\omega, \xi)$ at $M_1(2D)$, field in direction of negative mass ($a_{\parallel} < 0$). Values of effective masses correspond to $a_x = 0.15 \times 10^{-3}$ eV^2 s^2 g^{-1}; $a_y = -0.26 \times 10^{-3}$ eV^2 s^2 g^{-1}. Full line is for $a_{\parallel}^{1/3}(e\xi)^{2/3} = -1.0 \times 10^{-2}$ eV, dotted line for $a_{\parallel}^{1/3}(e\xi)^{2/3} = -1.28 \times 10^{-2}$ eV, dashed line for $\xi = 0$. (c) $J_{cv}(\omega, \xi)$ at $M_1(2D)$, field in direction of positive mass ($a_{\parallel} > 0$). Values of a_x, a_y same as (b). Full line is for $a_{\parallel}^{1/3}(e\xi)^{2/3} = 0.85 \times 10^{-2}$ eV, dotted line for $a_{\parallel}^{1/3}(e\xi)^{2/3} = 1.07 \times 10^{-2}$ eV, dashed line for $\xi = 0$.

that if an M_1 c.p occurs at $k_c \neq 0$ then it is necessary to sum the contributions to the electro-optic effect arising from a number of equivalent c.p's (points of the star of $\mathbf{k}_c$). Since the applied field makes different angles with axes of the effective mass tensor so the contribution from each point of the star must be evaluated separately in order to obtain the total electro-optic effect. In addition the contribution from each separate c.p may depend upon the polarization $\boldsymbol{\alpha}$ of the e.m radiation [64] via the matrix element $|\boldsymbol{\alpha} \cdot \mathbf{p}_{cv}|^2$. In principle an analysis of the $\Delta\varepsilon_2(\omega, \xi)$ spectra obtained for different orientations of $\boldsymbol{\alpha}$ and $\boldsymbol{\xi}$ yields values for the effective mass components.

The foregoing description of the form of $\varepsilon_2(\omega, \xi)$ due to direct interband transitions has to be improved by introducing the various interaction mechanisms which limit the lifetime of the excited state. This can be done in a phenomenological manner [50, 65, 66] by adding an imaginary term to the energy which has the effect of rounding the singularity in the optic functions $F(x)$ and $G(x)$, (98), and damping the subsidiary oscillations, Figures 11 and 12. When the damping is sufficiently large the subsidiary oscillations disappear leaving only the structure in the vicinity of the critical point. This situation seems to be characteristic of most high energy interband electro-reflectance spectra.

5.2. Forbidden interband transitions

Equations (99, 102) give the form of $\varepsilon_2(\omega, \xi)$ near three and two-dimensional c.p's for direct allowed transitions in which the matrix element $|\boldsymbol{\alpha} \cdot \mathbf{p}_{cv}|$ is assumed constant i.e. independent of $\mathbf{k}$. In the case of forbidden transitions, (29) the matrix element for transitions between initial state i and final state f is given by

$$P_{if} = \hbar |\nabla_{\mathbf{r}_\alpha} \phi(0)| C_1 \delta_{\mathbf{k}_i, \mathbf{k}_f}, \tag{103}$$

where C_1, independent of $\mathbf{k}$, involves the matrix elements between the periodic parts of the Bloch states at the band edges and $\phi(0)$ is the solution of (91). Assuming isotropic energy bands (M_0 c.p) and $\boldsymbol{\xi}$ in the z direction say the solution of (91) gives [49]

$$\varepsilon_2(\omega, \xi)_\| = S\theta^{3/2} \int_{(\omega_g - \omega)/\theta}^{\infty} |Ai'(t)|^2 \, dt \tag{104a}$$

$$\varepsilon_2(\omega, \xi)_\perp = \frac{S\theta^{3/2}}{2} \int_{(\omega_g - \omega)/\theta}^{\infty} \left(t + \frac{\omega_g - \omega}{\theta}\right) |Ai(t)|^2 \, dt \tag{104b}$$

for the two polarizations where θ is given by (93) and

$$S = \frac{2e^2\hbar}{m^2\omega^2} C_1 \left(\frac{2\mu}{\hbar}\right)^{5/2}.$$

When $\omega \ll \omega_g$ then (104) reduce to

$$\varepsilon_2(\omega, \xi)_{\|,\perp} = S\theta^{3/2} \exp\left[-\tfrac{4}{3}\frac{(\omega_g - \omega)^{3/2}}{\theta}\right] \quad (105)$$

giving an exponential tail in the energy gap similar to that for allowed M_0 transitions, Figure 11.

As in the field-free case structure in $\Delta\varepsilon_2(\omega, \xi)$ due to forbidden (weak) transitions at energies above threshold is likely to be masked by absorption due to neighbouring allowed transitions. Equations giving the behaviour of $\Delta\varepsilon_2(\omega, \xi)$ due to forbidden transitions at (3D) M_1, M_2, M_3, c.p's have been derived in the case of parabolic bands [67].

5.3. Indirect Transitions

The effect of an electric field on the indirect absorption threshold has been calculated in the effective mass approximation [69] for a crystal having a single valence (conduction) band maximum (minimum) at $\mathbf{k} = 0$ ($\mathbf{K}_0$) and a single phonon branch. Depending upon whether the transition involves creation (+) or destruction (−) of a phonon (frequency ω_{ph}, wave vector $\mathbf{K}_0$) the energy equation is

$$\hbar\omega \pm \hbar\omega_{ph} = E_f - E_i = E_g + W_1 + E_1, \quad (106)$$

where W_1 is the eigenvalue of (91) for ξ in the z direction and

$$\frac{2E_1}{\hbar^2} = \mathbf{K} \cdot M^{-1} \cdot \mathbf{K} + \frac{k_y^2}{\mu_y} + \frac{k_z^2}{\mu_z}, \quad (107)$$

where $\mathbf{K} = (\mathbf{k}_e + \mathbf{k}_h - \mathbf{K}_0)$ and $\mathbf{k} = \tfrac{1}{2}(\mathbf{k}_e - \mathbf{k}_h - \mathbf{K}_0)$.

The approximate solution of (91) at $r = 0$ is

$$\phi(0) = \frac{(2\mu_z\hbar)^{1/4}}{\hbar\pi^{1/2}\theta_0^{1/4}} Ai(t), \quad (108a)$$

where $\theta_0^3 = e^2|\xi|^2/2\mu_z\hbar$ and

$$t = (-W_1 + \hbar^2 k_x^2/2\mu_x + \hbar^2 k_y^2/2\mu_y)\hbar\theta_0. \quad (108b)$$

Substituting $\phi(0)$ in the general expression for indirect transitions gives finally

$$\frac{\omega^2}{D\theta^2}\varepsilon_2(\omega, \xi)_\pm = \frac{2^{2/3}}{32}\left\{Ai(r) + rAi'(r) + r^2\left[\tfrac{1}{3} - \int_0^r Ai(x)\,dx\right]\right\}, \quad (109)$$

where

$$D = 8e^2C^2(n_{\mathbf{K}_0} + \tfrac{1}{2} \pm \tfrac{1}{2})(m_e^* m_h^*)^{3/2}/\pi^2 m^2\hbar^4.$$

C^2 is a constant of proportionality, $n_{\mathbf{K}_0}$ is the number of phonons of wave vector $\mathbf{K}_0$ and $r = 2^{2/3}(E_g - \hbar\omega \pm \hbar\omega_{ph})/\hbar\theta_0 = 2^{2/3} V_0$ (which defines V_0). In the limit as

$\xi \to 0$ (109) gives the zero field situation (35). At the absorption edge (109) predicts that $\varepsilon_2(\omega, \xi) \propto \xi^{4/3}$ whereas far below the edge $\varepsilon_2(\omega, \xi) \propto V_0^{-7/4} \exp(-\frac{4}{3} V_0^{3/2})$. Similar expressions for $\varepsilon_2(\omega, \xi)$ have been obtained by time dependent perturbation methods [56, 70, 71] which also predict an oscillatory structure ('Stark ladder'), this vanishes however when electron collisions are taken into account.

5.4. Excitonic Transitions

When the Coulomb attraction between the electron and hole is included the effective mass equation, (91), becomes, for isotropic bands

$$\left(\frac{-\hbar^2}{2\mu}\nabla^2 - V(\mathbf{r})\right)\phi_n(\mathbf{r}) = \left(E_n - \frac{\hbar^2 K^2}{2M}\right)\phi_n(\mathbf{r}), \tag{110}$$

where

$$V(\mathbf{r}) = \frac{-e^2}{\varepsilon r} - e\mathbf{\xi} \cdot \mathbf{r} \quad \text{and} \quad \mathbf{r} = \mathbf{r}_e - \mathbf{r}_h.$$

The potential $V(\mathbf{r})$ is shown in Figure 13 for zero and finite fields $\mathbf{\xi}$. The applied field has the effect of lowering a lip on the Coulomb potential well which causes the bound (exciton) levels to be mixed and broadened into a continuum. The electric field will ionize the exciton if it provides a potential drop of at least 1 Ry across the effective Bohr radius; this ionization field is defined as [72]

$$\xi_I = R/ea = (\mu^2/m^2)\varepsilon^{-3} 2.59 \times 10^9 \text{ V cm}^{-1}. \tag{111}$$

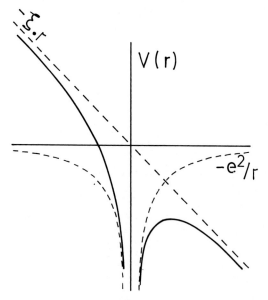

Fig. 13. The potentials $-\mathbf{\xi} \cdot \mathbf{r}$ and $-e^2/r$ (dashed line) together with the combination $-\mathbf{\xi} \cdot \mathbf{r} - e^2/r$ (full line).

In addition the applied field causes a slight widening of the potential well which produces a shift of the 1s level to lower energies – the second order Stark shift.

Expressions for $\varepsilon_2(\omega, \xi)$ obtained from (110) and the time dependent Schrödinger equation have shown [73] that in the case of a *weak* field and spherical energy bands the direct exciton peak positions are located at

$$-S_i = \frac{E_g - R/n^2 - \hbar\omega}{\hbar\Lambda_H}, \tag{112}$$

where $-S_i$ defines the maximum of the Airy function $Ai(x)$ occurring in the expression for $\varepsilon_2(\omega, \xi)$. The factor Λ_H in (112) is given by

$$\Lambda_H^3 = \theta_0^3 - (e^2 \xi \xi_H)/2\mu_x \hbar, \tag{113}$$

where ξ_H is of the order of the electron charge e divided by the Bohr radius and θ_0^3 is as in (99) except that now the reduced mass μ_x is that in the x direction. When $|\xi| \ll |\xi_H|$ then (112, 113) combine to give

$$\hbar\omega \simeq E_g - \frac{R}{n^2} - \hbar\left(\frac{e^2}{2\mu\hbar}\right)^{1/3}(\xi\xi_H)^{1/3}\left(1 - \frac{\xi}{3\xi_H}\right). \tag{114}$$

As ξ increases the peak position of the line is shifted first to lower and then to higher energies while the amplitude of the peak decreases. When $|\xi| > |\xi_H|$ then the $e-h$ interaction can be neglected and the expression for $\varepsilon_2(\omega, \xi)$ reduces to that for direct allowed transitions.

For the case when ξ is of the order of ξ_I (110) has been solved by direct numerical integration. Such a procedure has been used to determine the field dependence of the shift and broadening of the 1s exciton level [74] and exciton absorption tail [75], the latter calculation showing that at small ξ the absorption coefficient below the zero field (M_0) threshold is given by $\exp(-C|A-E|/f)$ where C, A are constants, $E = \hbar\omega - E_g$ and $f = e\xi a/R$.

A similar procedure has been used [72] to find the effect of an electric field on the absorption resulting from direct allowed transitions into delocalized exciton states and continuum states near M_0 and M_3 edges. As before, (63a), $\varepsilon_2(\omega) \propto |\phi_n(0)|^2$ where $\phi_n(\mathbf{r})$ satisfies (110) with $\mathbf{K} = 0$ since direct transitions are being considered, (59). The results of the calculation are given in Figure 14 in terms of a dimensionless density of states function $\phi^2(0)$ defined as

$$\phi^2(0) = 4\pi^2 a^3 \sum_n |\phi_n(0)|^2 \delta(E_g + E_n - \hbar\omega)/R. \tag{115}$$

Values of $\phi^2(0)$ are plotted for different values of the ratio $F = \xi/\xi_I$. For $F = 0.005$ ξ has little effect on the 1s or 2s exciton levels but does effect the $n = 3$ and higher levels; the $n = 3$ level is split into three parts and the higher levels are smeared into a continuum as a result of deformation of the Coulomb potential well, Figure 13. The three Stark split branches of the $n = 3$ level correspond to mixtures of the 3s, 3p and 3d hydrogenic states for small F, the lowest branch is

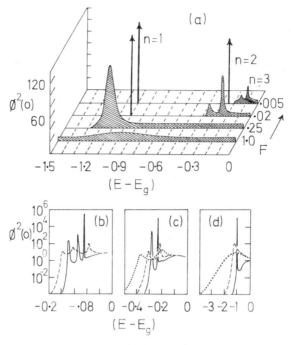

Fig. 14. [72] (a) Electric field effects on bound (Wannier) exciton levels for $F = \xi/\xi_I = 0.005$, 0.02, 0.25 and 1.0. The bound states are split and subsequently mixed into a continuum as the electric field lowers the lip of the Coulomb well – see Figure 13. Electric field effects on the (b) $n = 3$ hydrogenic level for $F = 0.0025$ (full line) and 0.0040 (dashed line). (c) $n = 2$ hydrogenic level for $F = 0.01$ (full line), 0.016 (dashed lines) and 0.025 (dotted line). (d) $n = 1$ hydrogenic level for $F = 0.10$ (full line) 0.32 (dashed line) and 1.0 (dotted line).

the one most broadened by ξ because [72] the associated wave function is concentrated on the 'lip' side of the potential well. Increasing F to 0.02 smears the $n = 3$ level into the continuum and splits the $n = 2$ level into two parts corresponding to mixtures of the 2s and 2p hydrogenic states, again the lowest energy branch is most broadened. Increasing F from 0.02 to 0.25, Figure 14, smears the $n = 2$ levels into the continuum and broadens and shifts the $n = 1$ level. This $n = 1$ exciton state shows the quadratic Stark shift to lower energies for $F < 0.5$, reaching a minimum energy in the range 0.4 to 0.5, and at $F \simeq 0.5$ begins to move back to higher energies, c.f (114), due to mixing with the continuum. For $F > 1.0$ no bound levels are distinguishable since the electric field has ionized the exciton, (111), and only the continuum states are important. At these larger F values solutions to (110) can be found for isotropic positive (M_0) and negative (M_3) reduced mass values, the associated $\phi^2(0)$ spectra resembling the zero electron-hole interaction spectra of Figure 11.

Thermally broadened versions of Figure 14 have been fitted to experimental $\Delta\varepsilon_2(\omega, \xi)$ spectra, see PbI_2 for example [72], any mismatch being attributed to non-uniformity of the applied field or the use of a Coulomb potential to describe the electron-hole interaction. If the electron-hole interaction is of a Koster-Slater type, Section 3, instead of Coulombic then the contact potential V is defined by

$$\langle R_j | V | R_{j'} \rangle = \tilde{g} \, \delta_{R_j,0} \, \delta_{R_{j'},0}, \tag{116}$$

where $|R_j\rangle$ is a Wannier function centred on lattice site R_j. In this case a perturbation calculation [76] shows that the interband dielectric function $\tilde{\varepsilon}$ of the solid is given by

$$\tilde{\varepsilon} = \tilde{\varepsilon}_0 (1 + g\tilde{\varepsilon}_0)^{-1}, \tag{117}$$

where g is related to the strength factor $\tilde{g}$ by $g = \omega^2 m^2 \tilde{g}/4\pi^2 e^2$ and $\tilde{\varepsilon}$ is the dielectric function when the solid has no direct electron-hole interaction but is perturbed by an applied electric field. If the perturbation is sufficiently small that $\tilde{\varepsilon}_0$ is related to the dielectric function ε_0 of the completely unperturbed solid by $\Delta\varepsilon_0 = \tilde{\varepsilon}_0 - \varepsilon_0$ then, (117), to first order the contact interaction V modifies the response to

$$\Delta\varepsilon = (1 + g\varepsilon_0)^{-2} \Delta\varepsilon_0. \tag{118}$$

Thus the Koster-Slater interaction mixes, via g, the real and imaginary parts of $\Delta\varepsilon_0$ and, in effect, varies the phase of $\Delta\varepsilon_0$. This result is the same as that obtained in Section 3.5 except that now it is the electric field (rather than zero field) line shapes associated with a particular critical point (c.p) that are relevant.

Consequently the conclusion is that g leads to a mixing of the M_t and M_{t+1} line shapes shown in Figure 11 so that the experimental line shapes cannot uniquely identify the type of c.p.

5.5. SYMMETRY ANALYSIS OF ELECTRO REFLECTANCE SPECTRA

The preceding sections have described the spectral line shapes to be expected for specific interband (or exciton modified) transitions. In principle comparison of the theoretical and experimentally observed electric field modulated line shapes should enable the transitions involved to be identified. In practice however the effective electric field in the sample can be non-uniform and even unknown so that direct comparison of experimental and theoretical line shapes is impossible. For these reasons the preferred experiment is that in which the change in reflectivity, $\Delta R/R$, say, is observed as the field orientation in the crystal varies. An analysis of the electric field and polarization dependence of the crystal reflectivity will often allow an unambiguous choice of the c.p symmetry [63, 64, 77].

6. The Effect of an Applied Magnetic Field on the Crystal Dielectric Function

The magneto-optic properties of a solid are primarily determined by the response of the electrons to the applied magnetic field. On the one-electron energy band model (i.e. excluding electron-hole interaction) two cases can be considered viz. simple bands and complex (i.e. degenerate) bands.

6.1. Simple energy bands

M_0 c.p (v.b maximum and c.b minimum at $\mathbf{k}=0$).
In the presence of a static, uniform magnetic field $\mathbf{H}$ the zero order conduction electron wave function is [78, 79]

$$\psi(\mathbf{r}) = u_0(\mathbf{r})F(\mathbf{r}), \tag{119}$$

where $u_0(\mathbf{r})$ is the periodic part of the Bloch function at the band extremum and $F(\mathbf{r})$ satisfies the effective mass equation

$$\frac{1}{2m^*}\left(\mathbf{p} + \frac{e}{c}\mathbf{A}\right)^2 F(\mathbf{r}) = E_j F(\mathbf{r}). \tag{120}$$

Here m^* is the effective mass at the extremum, E_j the energy of the electron state measured from the bottom of the unperturbed band (index j) and $\mathbf{A}$ is the vector potential of the applied magnetic field. When $\mathbf{H}$ is in the z direction the eigenvalues of (120) are

$$E_c = \hbar^2 k_z^2/2m_c^* + (n+\tfrac{1}{2})\hbar\omega_c, \quad n = 0, 1, 2 \text{ etc.}, \tag{121a}$$

where $\omega_c = eH/m_c^* c$.

Similarly the parabolic v.b splits into a series of sub-bands given by

$$E_v = -E_g - \hbar^2 k_z^2/2m_v^* - (n+\tfrac{1}{2})\hbar\omega_v. \tag{121b}$$

Each energy eigenvalue specified by (121) corresponds to a large number of degenerate states which differ from one another in the value of a third quantum number not appearing in the energy expressions. The density of states in each Landau level (specified n, k_z) is

$$\frac{d}{dk_z} N(n, k_z) = \frac{eH}{4\pi^2 c\hbar}. \tag{122}$$

$M_1 M_2$ c.p's (energy bands forming a saddle point).

With the origin at the c.p the energy can be written in the expanded form (c.f. (19))

$$E_k = E_0 + a_x \hbar^2 k_x^2 + a_y \hbar^2 k_y^2 + a_z \hbar^2 k_z^2, \tag{123}$$

where the axes are chosen so that a_x, a_y have the opposite sign to a_z. As before the energy values in the presence of a field $\mathbf{H}$ are found [80] by solving the effective mass equation (120). Defining

$$a = a_x a_y H_y^2 + a_z a_x H_y^2 + a_y a_z H_x^2 \tag{124a}$$

$$s = \mathbf{H} \cdot \mathbf{p} - (e/c) H_z H_y x \tag{124b}$$

$$b = a_y H_x^2 + a_x H_y^2 \tag{124c}$$

it is found that discrete quantum levels occur at

$$E - E_0 = (n+\tfrac{1}{2})\frac{b}{|b|}\frac{2e\hbar}{c}a^{1/2} + \frac{a_x a_y a_z}{a}s^2 \qquad (125)$$

when $a > 0$ i.e. the orientation of the magnetic field with respect to the z axis is within the elliptic cone $a > 0$. When $a < 0$ the Hamiltonian of (120) has continuous eigenvalues.

6.2. COMPLEX ENERGY BANDS

A much more complicated situation occurs when the unperturbed energy bands are degenerate. An illustration of this is the upper (Γ_8^+) spin orbit split v.b in *Ge* which is fourfold degenerate at $\mathbf{k} = 0$. An applied magnetic field splits the fourfold degenerate band into four sets of magnetic levels or ladders [79]. Unlike the Landau levels of (121) however the spacing of levels in a set is irregular at small n – the so called quantum effect. At large values of n the levels become equally spaced as in (121).

The parabolic c.b in *Ge* has two fold degeneracy due to spin and in this case the magnetic field energy levels are

$$E_c^{\pm}(n) = E_g + (n+\tfrac{1}{2})\hbar\omega_c \pm \tfrac{1}{2}\beta g_c H + \hbar^2 k_z^2/2m_c, \qquad (126)$$

where the effective g factor, g_c, for electrons is [81]

$$g_c = 2\left[1 + \left(1 - \frac{m}{m_c}\right)\frac{\Delta}{3E_g + 2\Delta}\right] \qquad (127)$$

and Δ is the spin-orbit splitting.

6.3. DELOCALIZED EXCITONS IN A MAGNETIC FIELD

6.3.1. *Parabolic (M_0) Excitons*

In the effective mass approximation the delocalized exciton can be treated as a two particle system with Coulomb attraction $e^2/\varepsilon |\mathbf{r}_e - \mathbf{r}_h|$ where $\mathbf{r}_e$, $\mathbf{r}_h$ are electron, hole coordinates. In a magnetic field [82]

$$\left[\frac{1}{2m_e}\left(\mathbf{p}_e + \frac{e}{c}\mathbf{A}_e\right)^2 + \frac{1}{2m_h}\left(\mathbf{p}_h - \frac{e}{c}\mathbf{A}_h\right)^2 - \frac{e^2}{\varepsilon |\mathbf{r}_e - \mathbf{r}_h|}\right]\psi = E\psi, \qquad (128)$$

where E is the energy relative to the bottom of the c.b and $\mathbf{A}_i = \tfrac{1}{2}(\mathbf{H} \wedge \mathbf{r}_i)$. Changing to centre of mass coordinates [83]

$$\mathbf{r} = \mathbf{r}_e - \mathbf{r}_h; \qquad \mathbf{R} = (m_e \mathbf{r}_e + m_h \mathbf{r}_h)/(m_e + m_h) \qquad (129a)$$

and introducing a wave function $\phi(\mathbf{r})$ for the relative motion of electron and hole defined by

$$\psi(\mathbf{r}_e, \mathbf{r}_h) = \phi(\mathbf{r})\exp\left[\frac{-ie}{2\hbar c}(\mathbf{H} \wedge \mathbf{R}) \cdot \mathbf{r}\right] \qquad (129b)$$

then (128) takes the simplified form [83]

$$\left[\frac{-\hbar^2\nabla^2}{2\mu} - \frac{ie\hbar}{2c}\left(\frac{1}{m_e} - \frac{1}{m_h}\right)\mathbf{H}\cdot\mathbf{r}\wedge\nabla + \frac{e^2}{8\mu c^2}(\mathbf{H}\wedge\mathbf{r})^2 - \frac{e^2}{\varepsilon r}\right]\phi_n(\mathbf{r})$$
$$= E_n\phi_n(\mathbf{r}), \quad (130)$$

where $\mu^{-1} = m_e^{-1} + m_h^{-1}$. In addition to the unperturbed Hamiltonian (130) contains the following magnetic terms
(a) the linear Zeeman term

$$\frac{ie\hbar}{c}\left(\frac{1}{m_e} - \frac{1}{m_h}\right)\mathbf{A}(\mathbf{r})\cdot\nabla = \frac{-e}{2c}\left(\frac{1}{m_e} - \frac{1}{m_h}\right)\mathbf{H}\cdot\mathbf{L}, \quad (131a)$$

where $\mathbf{L} = \mathbf{r}\wedge(-i\hbar\nabla)$ is the relative angular momentum operator. For the hydrogenic $p\pm$ states the wave functions have the form [84] $\psi_\pm \propto (x+iy)f(r) = f(r)\exp(i\phi)$ and treating (131a) as a perturbation it is found that the associated energy shift is

$$\Delta E_\pm = \frac{e\hbar}{2c}\left(\frac{1}{m_e} - \frac{1}{m_h}\right)H \quad (l=1, m_l = \pm 1) \quad (131b)$$

so that the s-states are unaffected by a magnetic field. The absence of an observed linear Zeeman shift in the exciton spectrum can be attributed to the near equality of m_e, m_h.
(b) a diamagnetic term which for $\mathbf{H}$ in the z direction becomes

$$\frac{e^2}{8\mu c^2}(\mathbf{H}\wedge\mathbf{r})^2 = \frac{e^2 H^2}{8\mu c^2}(x^2+y^2). \quad (131c)$$

Since (x^2+y^2) is proportional to the square of the exciton orbital radius so the predicted energy shift due to (131c) varies as exciton quantum number (n) to the fourth power, this behaviour is observed in MoS_2 [85] for example. Both s and p states are affected by (131c); for s states [86]

$$\Delta E_s = \frac{\varepsilon^2 \hbar^4}{8\mu^3 c^2 e^2} n^4 \quad (131d)$$

and for p states $\Delta E_p = 2\Delta E_s$.

In addition to terms (131a, c) the Hamiltonian of (131) should contain a term due to the centre of mass motion, c.f (40). Magnetic field effects due to this term are always present but are usually negligibly small.

Equation (128) cannot be solved exactly and various approximations have been employed depending upon the value of the parameter 2γ, defined as the ratio of the cyclotron precession energy $\hbar\omega_c(=\hbar eH/\mu c)$ to the Coulomb binding energy $R_e = \mu e^4/2\hbar^2\varepsilon^2$. In the weak field region $\gamma \ll 1$ the solution to (128) yields the energy levels (n, l, m) of a hydrogen atom [87, 88]. In the strong field limit solutions are obtained by neglecting the Coulomb term [82, 89], the eigenvalues of (128) are then

$$E^{NMk} = \frac{e\hbar}{c} H \left[\frac{1}{\mu}(N+\tfrac{1}{2}) - \frac{M}{m_h} \right] + \frac{\hbar^2 k^2}{2\mu}, \tag{132}$$

where $N = 0, 1, 2$ etc. and M takes integral values less than or equal to N. The first term in (132) is the usual cyclotron energy, the second arises from the angular momentum about the direction of **H** and the third is the free kinetic energy along that axis.

Over the intermediate field region calculations [90, 82, 91, 92, 93, 94] of the variation of state energy with γ support the 'non-crossing' rule [95], one consequence of this being that the zero-field $M = 0$ states lie, in the high field limit, below the lowest free electron Landau level.

Solutions to (128), in the high field limit, have been obtained [96] for **H** parallel to the optic axis of uniaxial crystals. The resulting eigenvalues involved the dielectric and reduced mass anisotropies which could be chosen to reproduce the observed [97] magnetic field dependence of excitons in the layer compound GaSe. The case of the two dimensional exciton in a magnetic field has also been treated [98, 99, 100] but is recognized as an inappropriate model for excitons in layer structures.

6.3.2. Hyperbolic Excitons

Assume that the non-degenerate electron and hole bands from which the exciton is formed have a common axis of symmetry (the z axis). When the angle θ between **H** and **z** satisfies the condition

$$\frac{1}{m_{e,h}^2(\theta)} = \frac{\cos^2\theta}{(m_{e,h}^2)_\perp} + \frac{\sin^2\theta}{(m_{e,h})_\perp (m_{e,h})_\parallel} > 0 \tag{133}$$

then the magnetic trajectories of the quasi particles lie in the vicinity of the hyperbolic point. Only this region of θ (known as the region of finite magnetic quantization, FMQ) will be considered.

Treating the more general case of crossed applied magnetic and electric fields and following a scheme similar to that used for spherical bands in which the exciton centre of mass motion (momentum **K**) is separated and then an adiabatic separation is made of the 'fast' magnetic motion and 'slow' Coulomb longitudinal motion it is found [101] that the kinetic energy is made up of (i) the centre of mass energy which depends on the electric field, (ii) the transverse Landau energy E_{n_1,n_2} where n_1, n_2 are the Landau quantum numbers in both bands and (iii) the longitudinal kinetic energy

$$T = \frac{1}{2\mu} \frac{\partial^2}{\partial v_3^2}, \tag{134}$$

where v_3 is the coordinate of longitudinal motion and

$$\frac{1}{\mu(\theta)} = \frac{m_e^2(\theta)}{(m_e^2)_\perp (m_e)_\parallel} + \frac{m_h^2(\theta)}{(m_h^2)_\perp (m_h)_\parallel}. \tag{135}$$

An electron hole bound state exists only in that region of θ (known as the discrete Coulomb quantization region, DCQ) where $\mu > 0$; the exciton ground state energy is then given by

$$E_L = -\frac{R_e}{L^2}, \tag{136}$$

where $R_e = 1/2 \mu a_0^2$, $a_0 = \varepsilon/\mu e^2$ and $1/L$ is a number, depending upon all the parameters, which diverges logarithmically for $(c\hbar/eH)^{1/2} = \lambda \ll a_0$, the Bohr radius.

On the boundary of the FMQ region (133) $\theta = \theta_f$ and the cyclotron mass $m_e^2 \to \infty$ or $m_h^2 \to \infty$; consequently $\mu \to 0$ and $E_L \to 0$. If the right hand side of

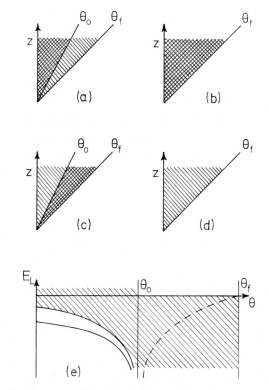

Fig. 15. Location of the regions of FMQ (single cross hatching) and DCQ (double cross hatching) for (a) $\mu_\parallel > 0$, $\theta_0 < \theta_f$; (b) $\mu_\parallel > 0$, $\theta_0 > \theta_f$; (c) $\mu_\parallel < 0$, $\theta_0 < \theta_f$; (d) $\mu_\parallel < 0$, $\theta_0 > \theta_f$; (e) the scheme of the absorption spectrum as a function of angle θ for case (a) above, energy is counted from the appropriate Landau level. Solid line denotes the lowest exciton band inside the DCQ region, shaded area denotes the continuous spectrum region, dashed line denotes maximum of continuous spectrum outside DCQ region [101].

(135) passes through zero for some angle θ_0 inside the FMQ region then the mass $\mu(\theta_0) \to \infty$. When $\theta \to \theta_0$ then $a_0 \to 0$ and $|E_L| \to \infty$ i.e. the level collapses. All possible cases are shown in Figure 15 where it is seen that the DCQ region may coincide with, or occupy some part of, the FMQ region or be completely absent. The DCQ region is adjacent to the (z) axis of symmetry when $\mu(0) = \mu_\| > 0$ and to the external surface of the FMQ cone when $\mu_\| < 0$. Considering these cases separately.

(a) $\mu_\| > 0$. If both $(m_e)_\perp$ and $(m_h)_\perp$ are positive then this situation can arise when the c.p in one band is type M_0 and in the other band is type M_1. If $(m_e)_\perp$ and $(m_h)_\perp$ have opposite signs this condition can occur from a conjunction of the c.p's $M_0 - M_3$ or $M_1 - M_2$.

(b) $\mu_\| < 0$. If $(m_e)_\perp$ and $(m_h)_\perp$ are both negative then this condition can occur from a conjunction of M_2, M_3 c.p's. However if $(m_e)_\perp / (m_h)_\perp < 0$ than a conjunction of M_0, M_3 or M_1, M_2 c.p's is necessary.

Referring to Figure 15 it follows that under the conditions of FMQ a range of angles θ can arise in which a discrete exciton spectrum occurs (and the continuous spectrum is that due to an attractive potential) and also an angular region in which the discrete spectrum is absent and the continuum corresponds to a repulsive potential. The change in the spectrum due to the rotation of $\mathbf{H}$ thus depends upon the numerical relation between the electron, hole parameters.

6.4. Intra and Interband Magneto Absorption in Semiconductors

Treating the e.m radiation as a perturbation an expression for the transition matrix element appropriate to a semiconductor having simple energy bands can be written using the zero order wave functions of (119) viz.

$$M_{if} = \int_{\text{crystal}} F_f^*(\mathbf{r}) \boldsymbol{\alpha} \left(\mathbf{p} + \frac{e}{c} \mathbf{A} \right) F_i(\mathbf{r}) \, d\tau \int_{\text{cell}} u_{fo}^* u_{io} \, d\tau$$
$$+ \int_{\text{crystal}} F_f^*(\mathbf{r}) F_i(\mathbf{r}) \, d\tau \int_{\text{cell}} u_{fo}^* \boldsymbol{\alpha} \cdot \mathbf{p} u_{io} \, d\tau, \quad (137)$$

where i, f refers to initial, final states and $F(\mathbf{r})$ and $\mathbf{A}$ (the vector potential of the periodic wave) are slowly varying compared with u_{fo}, u_{io}. The two terms of (137) account for intra and interband transitions respectively.

Intraband Transitions (Cyclotron resonance)

The periodic functions $u(\mathbf{r})$ of the valence (v) and conduction (c) bands are orthogonal so the first term in (137) is zero unless $u_{fo} = u_{io}$ i.e. transitions occur within the band in which case

(a) For $\mathbf{E} \perp \mathbf{H}$

$$M_{if} = \left(\frac{eH}{\hbar c} \right)^{1/2} \left(\frac{n+1}{2} \right)^{1/2} \quad (138a)$$

when the selection rules

$$\Delta k_y = \Delta k_z = 0 \quad \text{and} \quad \Delta n = \pm 1 \tag{138b}$$

are satisfied i.e. vertical transitions take place between neighbouring Landau levels of energy separation $\hbar\omega_c$.

(b) For $\mathbf{E} \parallel \mathbf{H}$

$$M_{if} = \hbar k_z \tag{139a}$$

when the selection rules are

$$\Delta k_y = \Delta k_z = 0 \quad \text{and} \quad \Delta n = 0. \tag{139b}$$

In this case 'transitions' are between states of the same energy so there is no photon absorption.

The effects of electron spin can be included by adding a term $g_{c(v)}\beta H M_J$ to (121) where $\beta = eH/2mc$; the effective g factor may differ considerably from the free electron value of 2 and M_J ($= \pm\frac{1}{2}$) is the component of angular momentum along $\mathbf{H}$. There is then an additional condition for allowed transitions viz. $\Delta M_J = 0$ for $\mathbf{E} \perp \mathbf{H}$, so that the transition is an electric dipole effect, and $\Delta M_J = 1$ for $\mathbf{E} \parallel \mathbf{H}$ in which case the transition takes place between the spin-split levels of *one* Landau state and is a magnetic dipole effect.

Interband Transitions (direct, allowed)

In this case u_{io} ($= u_{vo}$) and u_{fo} ($= u_{co}$) are orthogonal, the first term in (137) vanishes and the remaining term can be written

$$M_{vc} = (\boldsymbol{\alpha} \cdot \mathbf{p}_{cv}) \int F_c^*(\mathbf{r}) F_v(\mathbf{r}) \, d\tau, \tag{140}$$

where $\boldsymbol{\alpha} \cdot \mathbf{p}_{cv}$ is the momentum matrix element $\int u_{co}^* \boldsymbol{\alpha} \cdot \mathbf{p} u_{vo} \, d\tau$ and $\mathbf{p}_{cv}$ is non-zero when u_{co} and u_{vo} have opposite parity. The integral in (140) vanishes because of orthogonality unless the conditions

$$\Delta k_y = \Delta k_z = 0 \quad \text{and} \quad \Delta n = 0 \tag{141}$$

are satisfied. The transition energy is then given by (121a)–(121b). When electron spin is included a further selection rule $\Delta M_J = 0$ (for $\mathbf{E} \parallel \mathbf{H}$) and $\Delta M_J = \pm 1$ (for $\mathbf{E} \perp \mathbf{H}$) applies.

According to (141) allowed interband transitions only occur between Landau levels having the same quantum number n.

Including the density of such energy states per unit energy interval gives, neglecting spin, the absorption coefficient [102, 103] due to transition at an M_0 c.p viz.

$$\alpha_H = \frac{2e^2}{n_r \omega m^2 c} |\boldsymbol{\alpha} \cdot \mathbf{p}_{cv}|^2 \left(\frac{2\mu}{\hbar^2}\right)^{1/2} \left(\frac{eH}{\hbar c}\right) \sum_n [\hbar\omega - E_g - (n+\tfrac{1}{2})\hbar\omega_{cv}]^{-(1/2)}, \tag{142}$$

where $\omega_{cv} = \omega_c + \omega_v$ and n_r has been written for the refractive index to avoid confusion with the Landau level index n. Equation (142) shows that in a magnetic field the zero field M_0 absorption edge develops a series of absorption edges. The energy position of each peak plotted as a function of H gives a 'fan' set of straight lines (each corresponding to $n = n' = 0, 1, 2$ etc.) which intercept at $\hbar\omega = E_g$ and whose slopes gives the reduced mass μ. If electron spin effects are appreciable then each line will become a doublet whose splitting gives a measure of the combined g factor.

At saddle points in the energy band structure discrete quantum levels occur for a suitable orientation of the magnetic field, (124, 125), in which case, for a field in the z direction [104],

$$\varepsilon_2(\omega, H) = \frac{e^2}{m^2\omega^2} |\boldsymbol{\alpha} \cdot \mathbf{p}_{cv}|^2 \frac{eH}{\hbar c} \sum_n \int dk_z \, \delta\left(\pm|n+\tfrac{1}{2}|\hbar\omega_c \pm \frac{\hbar^2 k_z^2}{2|\mu_l^*|} - \hbar\omega + E_g\right),$$

(143)

where $(++) \to M_0$, $(+-) \to M_1$, $(-+) \to M_2$, $(--) \to M_3$.

μ_l^*, μ_t^* are the longitudinal ($\parallel \mathbf{H}$) and transverse reduced masses and $\omega_c = eH/|\mu_t^*|c$.

Indirect Transitions

The effect of an applied magnetic field on indirect transitions is mainly through a change in the density of states with the result that the absorption coefficient is given by [102, 103]

$$\alpha_H = 2K_\pm \left(\frac{eH}{\hbar c}\right)^2 \frac{(m_c m_v)^{1/2}}{\omega} \sum_{nn'} S(\hbar\omega - E_{nn'}), \qquad (144)$$

where

$$E_{nn'} = E_g + \hbar\omega_v(n'+\tfrac{1}{2}) + \hbar\omega_c(n+\tfrac{1}{2}) \pm \hbar\omega_p$$

and $S(x) = 0$ for $x < 0$ and $S(x) = 1$ for $x > 0$.

Since there is no selection rule on n and n' the indirect magneto absorption spectrum can be quite complicated.

6.5. Exciton Absorption in a Magnetic Field

6.5.1. Parabolic Excitons

As in the zero field case, (59), direct transitions to exciton states in a magnetic field take place only to pair states which have zero total momentum; the absorption is related to the matrix element of the momentum $\langle n, 0 | \mathbf{p} + (e/c)\mathbf{A} | 0 \rangle$ where quantum number n defines the relative electron-hole motion and the ket describes the ground state. Using the effective mass approximation the matrix element can be written

$$\langle n, 0| \boldsymbol{\alpha} \cdot \mathbf{p} + (e/c)\mathbf{A} |0\rangle = \left[\boldsymbol{\alpha} \cdot \mathbf{p}_{cv}(0) + \frac{ie\hbar H}{meE_g}(\boldsymbol{\alpha} \cdot \mathbf{p}_{cv}(0))\right]\varphi_n(0) +$$

$$[\mathbf{M}_{cv} + \boldsymbol{\pi}\varphi_n(\mathbf{r})]_{r=\infty}, \quad (145a)$$

where

$$\mathbf{M}_{cv} = \hbar^{-1}[\nabla_{\mathbf{k}}(\boldsymbol{\alpha} \cdot \mathbf{p}_{cv}(k))]_{\mathbf{k}=0} \quad (145b)$$

$$\boldsymbol{\pi} = -i\hbar\nabla + (e/2c)\mathbf{H} \wedge \mathbf{r} \quad (145c)$$

and φ_n is given by (130) which is appropriate to M_0 excitons.

In the case of allowed transitions only the first term in (145a) is required and hence $\varphi_n(0)$. Figure 16 shows the calculated [82] form of the allowed magneto

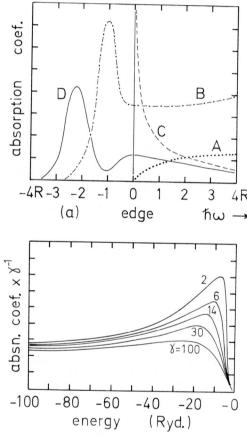

Fig. 16. (a) Comparison of the shapes of the allowed (M_0) absorption threshold for various cases [82]. (A) Free electron-hole pairs and (B) exciton effects with $H=0$. (C) Magnetic field, $\gamma=2$, but Coulomb effects neglected. (D) Magnetic field, $\gamma=2$, Coulomb effects included; a line width of ca. R_e was assumed for the exciton lines.

(b) Allowed absorption coefficient plotted against energy measured from the M_1 critical point for different values of the magnetic field [240].

Note that for higher magnetic fields the peak becomes stronger.

absorption threshold ($\gamma = 2$) for zero (curve C) and finite Coulomb attraction (curve D) between the electron and hole; the corresponding zero field thresholds, curves A and B respectively, are shown for comparison.

The Coulomb field has the effect of removing absorption intensity from the interband peak at the edge (associated with a Landau level) and concentrating this in the lowest exciton line (which increases in intensity and moves to lower energy with increasing field). As a result all exciton lines but the first merge with the edge forming a weak shoulder which peaks below the true edge.

Forbidden transitions come from the final term in (145a). For polarization $\boldsymbol{\alpha} \perp \mathbf{H}$ the form of the absorption is the same for left circularly polarized (transitions to $M = -1$ states) and right circularly polarized (transitions to $M = +1$ states) radiation except that the latter occurs at an energy lower by an amount $e\hbar H(m_h^{-1} - m_e^{-1})/c$. Expressions can also be derived for the pseudo and true continuum absorption.

6.5.2. Hyperbolic Excitons

The conditions under which a discrete hyperbolic exciton spectrum occurs in a magnetic field were described in Section 5.3; Figure 16b shows the way in which Coulomb interaction modifies the interband absorption edge.

The change in appearance of the hyperbolic exciton spectrum with orientation of **H** can be followed [101] by referring to Figure 15. For $\mu_0 > 0$ a Balmer series should occur, (136), but due to the broadening of the higher lines at hyperbolic points the spectrum will consist, in the DCQ region, of an isolated band at energy $E = -R_e/L^2$, $L \ll 1$, with a quasi continuous spectrum at $E > R_e$. As θ approaches θ_0 so μ_0 and R_e increase, also L increases logarithmically. Thus the isolated band and the continuous absorption edge are shifted to larger $|E|$, (taking R_e as an energy unit they will converge). At $\theta \simeq \theta_0$ the adiabatic condition, and hence the derived formulae, is invalid but the entire absorption will probably occur at $E < 0$ with a maximum near $E = -R_e \ln^2 (a_0/\lambda)^2$. With further change in θ the maximum moves to smaller $|E|$.

7. Combined Electric and Magnetic Fields

In many cases the information to be gained from magneto-optical measurements can be improved by applying a supplementary electric field to the sample. Following usual modulation procedures an alternating (superimposed on a static) electric field is used together with phase sensitive detection of the resulting differential absorption. The (alternating) electric and static magnetic fields applied to the sample are usually arranged to be mutually orthogonal or parallel to the incident light; the combined effects of these fields on the properties of the crystal are as follows.

7.1. Semiconductor having simple energy bands

The Schrödinger equation for electrons and holes in crossed (i.e. perpendicular) fields can be solved exactly for non-degenerate bands and from the eigenfunctions obtained the optical absorption can be derived as in the case of a magnetic field alone. The main results of such an analysis [105, 106] are:

(a) The selection rule $\Delta n = 0$, (141) valid for zero electric field ($\xi = 0$) breaks down in an electric field and transitions corresponding to $\Delta n = \pm 1, \pm 2 \cdots$ etc. have a finite transition probability.

(b) All transitions are shifted to lower photon energies by an amount

$$\Delta E = \left(\frac{m_e + m_h}{2}\right) \frac{\xi^2}{c^2 H^2}. \tag{146}$$

A similar result is obtained [107] for optical transitions at saddle points. Combining the measured values of ΔE with a value for the reduced mass $\mu = m_e m_h/(m_e + m_h)$ obtained from the ordinary magneto absorption spectrum, (142), allows m_e, m_h to be determined.

When the applied electric and magnetic fields are parallel then, within the effective mass approximation, the equations of motion of electrons and holes are separable since the effects of parallel fields are independent. The general result [108, 109] is that the absorption edges and peaks are broadened due to field induced tunnelling between the bands, i.e. the electric field has the effect of smearing out the Landau levels so bringing the $\varepsilon_1(\omega)$, $\varepsilon_2(\omega)$ spectra back to the zero magnetic field form [110].

7.2. Semiconductors having complex energy bands

The effect of crossed electric and magnetic fields on semiconductors having complex energy bands can be determined by a perturbation treatment [106] provided the $\xi = 0$ wave functions and energy levels are known. Such an analysis shows that the intensity of allowed transitions ($\Delta n = 0, -2$) is reduced while formerly forbidden transitions ($\Delta n = \pm 1, -3$) now obtain a finite transition probability. A further result is that the coupling between the valence band ladders is weak so that, to a good approximation, each ladder behaves, under the influence of an electric field, almost as a simple band. Since, for a simple band, the transition probabilities are proportional to $(m_e + m_h)^2$ so the heavy hole transitions should show more strongly in the differential spectrum (and be identified thereby) than the light hole transitions.

7.3. Indirect transitions

When a crossed electric field ξ is applied the onset of the phonon assisted magneto absorption is shifted from its $\xi = 0$ position by an amount.

$$\Delta E = \tfrac{1}{2}(m_c + m_v) V_H^2 - \hbar q_0 V_H, \tag{147}$$

where $V_H = c\xi/H$ is the nett electron drift velocity. The first term arises from the field dependence of the energy gap and the second (linear) term is due to the Doppler shift in frequency of the phonon involved in the transition, the shift in position of the absorption step being the same for either phonon absorption or emission.

7.4. Exciton transitions

In the crossed field configuration the (delocalized) exciton series limits (the Landau levels) are shifted to lower energies by the applied electric field, the exciton ionization energy decreasing as ξ increases. The ξ modulated exciton absorption spectrum can be used to determine the total exciton mass and so identify the lines associated with the light and heavy holes.

8. Stress Modulated Spectra

Hydrostatic stress applied to a crystal changes the interatomic spacings but not the crystal symmetry. As a result the absorption spectrum of a crystal under hydrostatic stress is similar (ignoring broadening) to the temperature modulated spectrum except that the different stress coefficients of the various energy gaps make certain transitions more, or less, easily detectable.

Uniaxial stress applied to a crystal changes the interatomic spacing and, in many cases, the crystal symmetry. This can result in the removal of energy degeneracies with the subsequent identification of transition symmetries and determination of deformation potentials. The majority of the uniaxial stress measurements have been made on the diamond, zincblende and wurtzite semiconductors and theoretical analysis of the effects of applied stress on the electron energy band structure has been mainly concerned with these materials.

In practice it is the stress modulated reflectivity (or transmissivity) $\Delta R/R$ that is measured which, together with the associated phase angle modulation $\Delta\theta$ (measured or obtained from $\Delta R/R$ by a dispersion relation) and the static parameters R, θ, completely specifies the system. For cubic crystals $\Delta R/R$ and $\Delta\theta$ are second rank tensors (when measured with polarized light in the principal directions of the perturbed crystal) each having three independent components. For normal incidence the functions $\Delta\varepsilon_1(\omega)$, $\Delta\varepsilon_2(\omega)$ are given by

$$\Delta\varepsilon_1 = \tfrac{1}{2}[n(\varepsilon_1-1)+\varkappa\varepsilon_2]\,\Delta R/R + [\varkappa(\varepsilon_1-1)+n\varepsilon_2]\,\Delta\theta, \tag{148a}$$

$$\Delta\varepsilon_2 = \tfrac{1}{2}[\varkappa(\varepsilon_1-1)+n\varepsilon_2]\,\Delta R/R - [n(\varepsilon_1-1)-\varkappa\varepsilon_2]\,\Delta\theta. \tag{148b}$$

It is the structure in $\Delta\varepsilon_2$ which can be related to the symmetry dependent shifts of the critical points and the strain dependent transition matrix elements. In a cubic crystal the fourth rank tensor $\mathbf{W}_{ijkl}$ relating the strain tensor to the dielectric tensor [113]

$$\Delta\hat{\varepsilon}_{ij} = (\Delta\varepsilon_1 + i\Delta\varepsilon_2)_{ij} = (W_1 + iW_2)_{ijkl}e_{kl} \tag{149}$$

reduces to the six independent components $(W_1)_{11}$, $(W_2)_{11}$, $(W_1)_{12}$, $(W_2)_{12}$, $(W_1)_{44}$, $(W_2)_{44}$ which completely describe the piezo-optical action on $\hat{\varepsilon}$. These components can be determined from properly chosen experiments, two directions of stress with light polarized parallel and perpendicular to each direction are sufficient. Thus with strain e along a cube edge [001] (149) reduces to

$$\Delta\varepsilon_\| = \left[W_{11} + 2\frac{S_{12}}{S_{11}} W_{12} \right] e, \tag{150a}$$

$$\Delta\varepsilon_\perp = \left[\frac{S_{12}}{S_{11}} W_{11} + \left(1 + \frac{S_{12}}{S_{11}}\right) W_{12} \right] e, \tag{150b}$$

where S_{11}, S_{12}, S_{44} are the wavelength independent cubic elastic compliances and $\Delta\varepsilon$ are the changes in the dielectric constant for light polarized parallel ($\|$) and perpendicular ($\perp$) to the stress direction. Stress diagonally along a face in the direction [011] gives

$$\Delta\varepsilon_\| = \left[\frac{(S_{12}+S_{11})W_{11} + (S_{11}+3S_{12})W_{12} + S_{44}W_{44}}{S_{11}+S_{12}+\frac{1}{2}S_{44}} \right] e, \tag{151a}$$

$$\Delta\varepsilon_\perp = \left[\frac{(S_{11}+S_{12})W_{11} + (S_{11}+3S_{12})W_{12} - S_{44}W_{44}}{S_{11}+S_{12}+\frac{1}{2}S_{44}} \right] e. \tag{151b}$$

Measurements (150, 151) give W_{11}, W_{12}, W_{44}. In the simplest case of non-degenerate bands and strain independent transition matrix elements the equality (or zero value) of some, or all, of the tensor components can be used to locate the band edge within the Brillouin zone.

9. The Effect of Temperature on the Crystal Dielectric Function

The atomic vibrations (phonons) present in a crystal at a given temperature can (a) allow indirect interband and excitonic transitions to occur, 2.2, 3.6, 7 (b) produce a shift in energy and broadening of the structure due to specific rigid lattice interband excitonic transitions, (c) broaden the free carrier plasma resonance structure.

In order to reduce the effects of thermal broadening optical reflection, transmission measurements are usually made on samples at low temperature. This often sharpens structure in the spectrum, particularly at the absorption threshold, by eliminating the phonon absorption component of the indirect transitions and the phonon absorption process (only) which broadens electron (hole) states at the lowest (highest) c.b (v.b) minima (maxima). This temperature dependence allows temperature modulation to be used to enhance structure present in the crystal spectrum at a given temperature.

The temperature modulated reflection spectrum is usually referred to as thermo reflectance spectrum.

9.1. Temperature Modulated Indirect Transition Spectra

The optical absorption due to indirect (phonon assisted) transitions is given by (35), the absorption intensity being proportional to N_p $\{= \exp(E_p/kT) - 1\}$ for phonon absorption processes and proportional to (N_p+1) for phonon emission processes. Assuming that the temperature dependence of N_p is more important than the temperature variation of $E_{g'}$ and the associated changes in the density of states then differentiating (35) gives [114]

$$\frac{d}{dT}\varepsilon_2 = \varepsilon_2(N_p+1)\frac{E_p}{kT^2}, \qquad \hbar\omega > E'_g - E_p \qquad (152a)$$

$$= 0, \qquad \hbar\omega \leq E'_g - E_p \qquad (152b)$$

for phonon absorption processes and

$$\frac{d}{dT}\varepsilon_2 = \varepsilon_2 N_p \frac{E_p}{kT^2}, \qquad \hbar\omega > E'_g + E_p \qquad (153a)$$

$$= 0, \qquad \hbar\omega \leq E_{g'} + E_p \qquad (153b)$$

for phonon emission processes. In both cases the temperature modulated spectrum is proportional to ε_2 and not to the frequency derivative of ε_2. Consequently there is no enhancement of the structure which characterizes the indirect spectrum.

Since $d(\ln \varepsilon_2)/dT = \varepsilon_2^{-1} d\varepsilon_2/dT$ so (152, 153) imply that $d(\ln \varepsilon_2)/dT$ should be either a constant or zero depending upon the frequency range considered.

For interband transitions temperature modulation of E'_g produces only weak structure in $d\varepsilon_2/dT$ due to the derivative of $(\hbar\omega - E'_g \pm E_p)$, (35). At low temperatures however excitonic effects can produce a strong $(\omega - \omega'_g)^{-(1/2)}$ singularity in the temperature modulated spectrum.

9.2. Temperature Modulated Direct Transition Spectra

Electron-phonon interaction in the crystal gives rise to a change in the band gap energy E_g (i.e. the c.p frequency $\omega_{cv}(\mathbf{k}_0)$, (16)) and introduces a broadening parameter γ. The effect on the temperature modulated spectrum of the change in E_g (typically 5×10^{-4} eV/°C) is usually greater than that due to the temperature derivative of γ which is $\simeq k$ ($= 10^{-4}$ eV/°C). Considering only the temperature dependence of E_g the derivative of the thermally broadened form of (13) gives the following expressions for the change $\Delta\varepsilon$ due to a temperature change ΔT [115]

$$\Delta\varepsilon_1 = A\frac{d}{dT}E_g \Delta T \int_0^\infty \frac{(E'-E)}{(E'-E)^2+\gamma^2}\frac{d}{dE_g}J_{cv}(E',E_g)\,dE', \qquad (154a)$$

$$\Delta\varepsilon_2 = A\frac{d}{dT}E_g \Delta T \int_0^\infty \frac{\gamma}{(E'-E)^2+\gamma^2}\frac{d}{dE_g}J_{cv}(E',E_g)\,dE', \qquad (154b)$$

where $J_{cv}(E', E_g)$ is the joint density of states appropriate to the particular c.p, (21–26). For $\gamma \ll E_g$ (154) can be given in closed form to a good approximation. The final expressions for the associated changes in the normal incidence reflectivity R are found by writing the total differential of R in terms of $\Delta\varepsilon_1$, $\Delta\varepsilon_2$ [116]

i.e.

$$\Delta R/R = \alpha(\omega)\,\Delta\varepsilon_1 + \beta(\omega)\,\Delta\varepsilon_2, \tag{155}$$

where α, β are frequency dependent coefficients involving n and $\varkappa$. Hence

(a) at an M_0 c.p

$$\frac{\Delta R}{R} = B\{-\alpha F(-x) + \beta F(x)\}\frac{d}{dT}E_g\,\Delta T; \tag{156a}$$

(b) at an M_1 c.p

$$\frac{\Delta R}{R} = B\{\alpha F(x) - \beta F(-x)\}\frac{d}{dT}E_g\,\Delta T; \tag{156b}$$

(c) at an M_2 c.p

$$\frac{\Delta R}{R} = B\{\alpha F(-x) + \beta F(x)\}\frac{d}{dT}E_g\,\Delta T; \tag{156c}$$

(d) at an M_3 c.p

$$\frac{\Delta R}{R} = B\{-\alpha F(x) + \beta F(-x)\}\frac{d}{dT}E_g\,\Delta T, \tag{156d}$$

where B is a constant, $x = (E - E_g)/\gamma$ and the function $F(x) = (x^2+1)^{-(1/2)}[(x^2+1)^{1/2} + x]^{1/2}$.

Since

$$\left(\frac{\partial\varepsilon}{\partial T}\right)_{\gamma\,\text{const}} = \frac{\partial\varepsilon}{\partial E_g}\frac{\partial}{\partial T}E_g = \frac{-\partial\varepsilon}{\partial E}\frac{\partial}{\partial T}E_g \tag{157}$$

so the temperature modulated $\Delta\varepsilon_1$, $\Delta\varepsilon_2$ line shapes at a singularity are essentially the same as those produced by wavelength modulation but multiplied by $-\partial E_g/\partial T$, where $\partial E_g/\partial T$ is usually negative.

Similar considerations apply to temperature modulated exciton spectra. As with the associated band edge the temperature shift in exciton energy usually outweighs the change in broadening parameter and is mainly responsible for the temperature modulated form of the different exciton line shapes [117].

9.3. Temperature Modulated Plasma Resonance Spectra

The 'free carrier' contribution to the crystal dielectric function may be written as

$$\hat{\varepsilon} = \varepsilon_L\left[1 - \frac{\omega_p^2}{\omega(\omega \pm i\gamma)}\right], \tag{158}$$

where $\omega_p = [4\pi Ne^2/m^*]^{1/2}$ is the plasma frequency and ε_L represents the background dielectric constant. Introducing parameters $W = \omega/\omega_p$ and $\eta = \gamma/\omega_p$ (158)

can be written, to first order in η/W as [118]

$$\varepsilon_1 = \varepsilon_L[1 - 1/W^2], \qquad \varepsilon_2 = \eta\varepsilon_L/W^3. \tag{159}$$

Temperature modulation of ε_1, ε_2 can occur as a result of change in ω_p (due to change in N or m^* with temperature) or broadening parameter γ. The change in normal incidence reflectivity associated with the temperature induced changes $\Delta\varepsilon_1$, $\Delta\varepsilon_2$ is obtained from (155, 158) viz.

$$\frac{1}{R}\frac{dR}{dW} = \frac{\delta}{W^3}\varepsilon_L\{2n(n^2 - 3\varkappa^2 - 1) - 3(\eta/W)\varkappa(3n^2 - \varkappa^2 - 1)\}, \tag{160}$$

$$\frac{1}{R}\frac{dR}{d\eta} = \frac{\delta}{W^3}\varepsilon_L\{\varkappa(3n^2 - \varkappa^2 - 1)\}, \tag{161}$$

where $\delta/2 = (n+\varkappa)^{-2}[(n+1)^2 + \varkappa^2]^{-2}$.

The derivative of the reflectivity with respect to the frequency ω or plasma frequency ω_p is obtained from the function $(1/R)(dR/dW)$, (160), shown plotted in Figure 17 for $\eta = 0.3$. This function had a negative peak close to the plasma

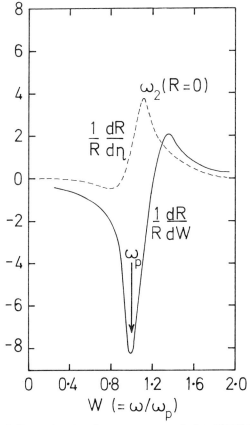

Fig. 17. Thermal reflectivity spectra at a plasma resonance calculated [118] for $\eta = 0.3$, (160, 161). $(1/R)(dR/dw)$ corresponds to plasma frequency modulation and $(1/R)dR/d\eta$ to broadening modulation.

frequency, $W=1$, the peak sharpening with decreasing η. The derivative of R with respect to the broadening parameter is also plotted in Figure 17; this function has a positive peak close to the reflectivity minimum ($dR/dW=0$) and so strongly modulates the reflectivity in this region. The thermoreflectance spectrum therefore consists of two peaks, a negative peak at frequency $\omega_1 \simeq \omega_p$ and a positive peak close to $R_{\min}(\omega_2)$, where $\varepsilon_1 = 1$, giving, (159), $\varepsilon_2 \simeq [1-(\omega_p^2/\omega^2)]^{-1}$.

10. The Measured Optical Properties of Layer Compounds

Most of the solids occurring as layer crystals fall into the following classification –
   (a) Halides of the Groups II, IV, V and transition metal elements.
   (b) Chalcogenides of the Groups III, IV, V and transition metal elements.
   (c) Graphite.
   (d) Micas.
   (e) Metal hydroxides.

Optical measurements have been made on only a few of these layer materials viz. those where it has proved possible to obtain suitably large stable crystals. The following sections describe the measurements that have been made on the most extensively studied crystals.

The ultimate aim of the optical measurements has been to establish the vibrational and electron energy band structure of the crystal. In the case of the electron energy band structure the optical measurements should, ideally, yield the $\varepsilon_2(\omega)$ spectrum of the crystal over a wide energy range to allow comparison with the $\varepsilon_2(\omega)$ spectrum derived from band structure calculations, (13). In practice the measurement of the crystal $\varepsilon_2(\omega)$ spectrum is a comparatively difficult and time consuming operation and the outstanding experimental advantage of the more flexible layer crystals is that (a) reliable reflectivity measurements can be made from the optically smooth crystal surface (b) thin crystals, obtained by successive cleaving using adhesive tape, can be prepared which allow transmission measurements even in the intrinsic absorption region. None the less $\varepsilon_2(\omega)$ spectra have been obtained only in a few cases and over a restricted energy range. The majority of the measured optical transmission, reflection spectra have been interpreted by assigning any observed structure to specific excitonic or interband transitions (utilizing any systematic trends within an isoelectronic series and any published band structures); in a few cases the nature of the transition has been established from field modulation spectra.

10.1. Group II dihalides

Cadmium Iodide
CdI_2 has the $Cd(OH)_2$ type crystal structure. The bonding within an I—Cd—I layer is predominantly ionic while the van der Waals type bonding between the layers allows easy cleavage in the basal plane. The crystal (space group D_{3d}^3) possesses a centre of inversion so that the infra red and Raman active modes are mutually exclusive, the Raman active phonons consisting of one A_{1g} and one

(lower frequency [121]) doubly degenerate E_g mode. The measured first order Raman line frequencies are at 113 [122], 111.5 [123], 112 [124] cm^{-1} attributed to the A_{1g} mode and at 32 [122], 44 [122], 43.5 [123], 28.2 [124] cm^{-1}.

A number of structural polytypes of CdI_2 occur, commonly identified by the number of iodide layers in the hexagonal unit cell [125] (e.g. 2H for the Cd(OH)$_2$ arrangement). Crystals grown from the melt and slowly from solution usually have the 4H structure [126, 127] but, as for all layer crystals showing polytypism, it is important to identify the type of crystal on which measurements are made.

Single crystals of CdI_2 are uniaxial negative [128] i.e. the ordinary refractive index n_0 for light having its electric vector **E** vibrating perpendicular to the crystallographic (optic) c axis of the crystal is greater than the extraordinary refractive index n_e appropriate to $\mathbf{E} \parallel \mathbf{c}$. The birefringence $\Delta = n_0 - n_e$ increases [128, 129] from ca. 0.2 at $\lambda = 0.65$ μm to ca. 0.3 near the absorption threshold at $\lambda = 0.4$ μm; this suggests that the absorption edge for $\mathbf{E} \parallel \mathbf{c}$ occurs to the high energy side of that for $\mathbf{E} \perp \mathbf{c}$. The absorption threshold, $\mathbf{E} \perp \mathbf{c}$, has the form appropriate to indirect transitions [130], measurements on the 4H polytype [131] giving an indirect gap of 3.266 eV at 300°K and a phonon energy of 0.034 eV (265 cm^{-1}).

Figure 18 shows the absorption spectrum of a thin CdI_2 film formed by

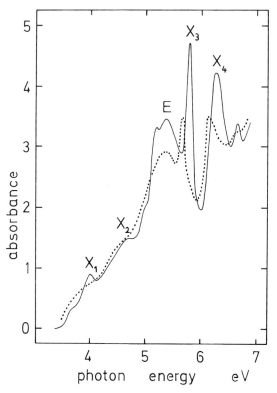

Fig. 18a. The absorption spectra of a thin CdI_2 film at room temperature (dotted curve) and liquid nitrogen temperature (full line curve) [132].

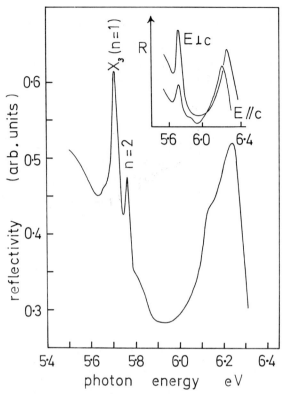

Fig. 18b. Reflectivity spectrum, $E \perp c$, $q \parallel c$, of a cleaved CdI_2 crystal (4 K). Insert shows the reflectivity spectrum of a CdI_2 surface (77 K) having the optic, c, axis lying in the plane of the surface [133].

evaporation onto a heated substrate [132] together with the single crystal reflection spectrum, $E \perp c$ [130, 133].

The most distinctive feature is the sharp ($n = 1$) exciton line, labelled X_3, at 5.683 eV (80°K) with the $n = 2$ quantum state at 5.761 eV and the series limit at 5.79 eV seen in the 4°K reflection spectrum. A further broader exciton line, labelled X_4, occurs at 6.195 eV; this line moves to 6.153 eV for $E \parallel c$. Peaks X_3, X_4 are generally identified with the halogen doublet which arises in the following way [134]. The lowest excited states of a halide ion having the inert gas configuration np^6 are $np^5 nd$, $np^5(n+1)s$ and $np^5(n+1)p$, the latter two states having almost the same energy. For a halide ion situated in the crystal electric field the $(n+1)p_z$ and $(n+1)s$ wave functions are hybridized, the p_x and p_y functions having lobes in the plane of the I^- layer. The separation of the two hybrid levels is 1–1.5 eV and both can be reached by allowed transitions from the ground state since both have appreciable s-character. Spin-orbit interaction splits the hybrid levels into doublets so that four exciton peaks should be observed. Peaks X_3, X_4 are identified with the high energy halogen doublet. Identification of the low energy doublet is less certain but comparison with other iodide spectra

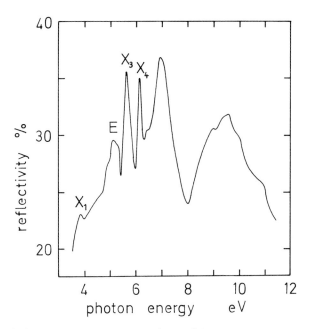

Fig. 18c. The single crystal reflection spectrum ($\mathbf{E} \perp \mathbf{c}, \mathbf{q} \| \mathbf{c}$) of CdI_2 at room temperature [130].

[134] (where the peak energy is expected to vary with interionic distance) assigns the low energy halogen doublet in CdI_2 to peaks labelled X_1, X_2 in Figure 18. None of the other CdI_2 absorption, reflection, peaks have been positively identified and no definite conclusions can be drawn about the form of the CdI_2 electron energy band structure.

Mercuric Iodide, HgI_2

At room temperature HgI_2 occurs as red tetragonal crystals (D_{4h}^{15}) which change to a yellow orthorhombic form on heating to 400°K [137] although, with care, the yellow form can be retained on cooling to room temperature.

For the tetragonal form group theory predicts four Raman active internal modes ($A_{1g} + B_{1g} + 2E_g$) and two Raman active translational modes ($B_{1g} + E_g$); the observed [135, 136, 140] Raman line frequencies (cm^{-1}) have been identified as follows [135, 136], $17(E_g)$, $29(E_g)$, $114(A_{1g})$, $142(B_{1g})$. The strong infra-red absorption line at 112 cm^{-1} [138, 139] is identified [136] as an E_u acoustic vibration mode.

Red tetragonal HgI_2 cleaves easily to give crystallographic planes perpendicular to the crystal c axis. Normal incidence reflectivity measurements, light propagation vector $\mathbf{q} \| \mathbf{c}$, give the ordinary reflection spectrum corresponding to the light electric field vector $\mathbf{E} \perp \mathbf{c}$. Reflectivity measurements from the 'as grown' crystal planes parallel to $\mathbf{c}$ give either the ordinary spectrum $\mathbf{E} \perp \mathbf{c}$, $\mathbf{q} \perp \mathbf{c}$ or the extraordinary spectrum $\mathbf{E} \| \mathbf{c}$, $\mathbf{q} \perp \mathbf{c}$. Figure 19 shows the $\mathbf{E} \perp \mathbf{c}$, $\mathbf{q} \| \mathbf{c}$ reflectivity spectrum [141] of red HgI_2 (4.2 K), the labelled structure occurs at A (2.339 eV), B

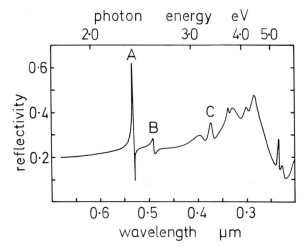

Fig. 19. The single crystal reflectivity spectrum of red HgI$_2$ (4.2 K) for $\mathbf{E}\perp\mathbf{c}, \mathbf{q}\|\mathbf{c}$ [141].

(2.538 eV), C (3.35 eV), peak A being at the absorption threshold. The sharp peak A is identified with an $n=1$ exciton absorption line having an oscillator strength, per unit cell, of 1.2×10^{-3} [142, 143, 144]; the $n=2$ member of the series is seen as a weak reflection peak at 2.368 eV, Figure 20, while the $n=1, 2, 3$ members of the hydrogen-like series have been observed in the absorption spectrum [145] giving an exciton binding energy of 0.029 eV. Electroabsorption (EA) and reflection (ER) measurements around A ($n=1$) [146, 147, 148] have not positively identified this exciton transition although the field dependence of the EA signal [148] was similar to that for indirect transitions.

The ordinary reflection spectrum, $\mathbf{E}\perp\mathbf{c}, \mathbf{q}\perp\mathbf{c}$ over region of peaks A and B, Figure 21, is similar to that for $\mathbf{E}\perp\mathbf{c}, \mathbf{q}\|\mathbf{c}$ although absorption measurements show an unexplained weakening of A for this classically equivalent situation

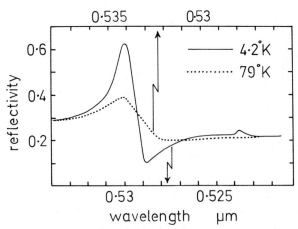

Fig. 20. Reflectivity spectra of red HgI$_2$ at 4.2 K (solid line) and at 79 K (dotted line – upper wavelength scale) [141] near the fundamental edge in enlarged scale, $\mathbf{E}\perp\mathbf{c}, \mathbf{q}\|\mathbf{c}$.

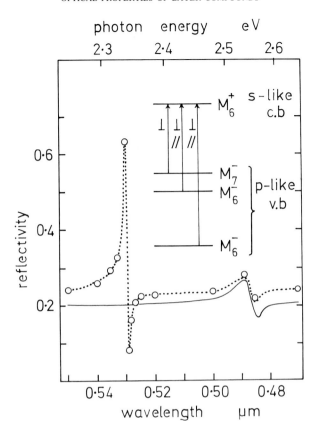

Fig. 21. Reflectivity spectra of HgI$_2$ (4.2 K) near the fundamental edge, [141], for different light polarizations. Dotted line–**E**⊥**c**, **q**⊥**c**. Circles–**E**⊥**c**, **q**∥**c**. Full line–**E**∥**c**, **q**⊥**c**. Inset shows a possible band structure for HgI$_2$ at the Γ point which accounts for the observed polarization dependence of reflection peaks A, B and C.

[142]. In the extraordinary reflection spectrum, Figure 21, exciton peak A is missing while B occurs in the same position as before. Peak C occurs in the same position for both the ordinary and extraordinary spectra.

The energy separation of peaks A and C (1.012 eV) is similar to the spin-orbit splitting observed in many iodides and close to the value of the spin-orbit splitting in atomic iodine (0.936 eV). This suggests [141] that the absorption peaks in (ionic) HgI$_2$ are due to electron transfer from the 5p states of the I$^-$ ion to the 6s states of the nearest neighbour Hg^{++} ions. On an energy band scheme, inset to Figure 21, this corresponds to Γ point transitions from a predominantly p-type valence band (v.b) to an s-type conduction band (c.b). The upper v.b state, coming mainly from the $j=\frac{3}{2}$ state of I$^-$, is split by the tetragonal crystal field into peaks A and B, the calculated selection rules (for **E**∥,⊥**c**) being consistent with experiment. Transitions from the lower spin-orbit split v.b level ($j=\frac{1}{2}$) to the c.b give rise to peak C. Structure in the reflection spectrum, Figure 19, to the high energy side of C has not been identified.

An ordinary and extraordinary emission spectrum of HgI_2 has also been recorded [149], the ordinary spectrum showing two emission lines shifted by 70 and 38 cm^{-1} to the low energy side of the first strong absorption line. These emission lines could respectively be due to bound exciton emission and radiative exciton recombination with emission of an LO phonon.

10.2. Group IV halides, $PbCl_2$, $PbBr_2$, PbI_2

Lead Chloride

$PbCl_2$ is isomorphous with $PbBr_2$ [150], both compounds crystallizing in the orthorhombic system (D_{2h}^{16}); the $PbCl_2$ structure can be viewed as a distorted close-packed array of halogen ions with lead ions accomodated in the same plane [134]. A factor group analysis of the primative cell predicts 18 Raman active (g) and 15 infra-red active (u) optical phonons divided into modes of the following symmetry.

$$\Gamma_{vib} = 6A_g + 6B_{1g} + 3B_{2g} + 3B_{3g} + 3A_\mu + 2B_{1\mu} + 5B_{2\mu} + 5B_{3\mu}$$

which excludes the three rigid translations of the whole unit cell. Lines in the Raman scattering spectrum (range 10–210 cm^{-1}) of an argon laser irradiated single crystal have been assigned [151] using the approximation that (a) modes involving motion mainly of the lead atoms should occur at similar frequencies (<100 cm^{-1}) in $PbCl_2$ and $PbBr_2$. (b) Modes involving motion mainly of the chlorine atoms should occur at frequencies greater than 100 cm^{-1} and be shifted by a factor (mass of Cl mass^{-1} of Br)$^{1/2}$ in $PbBr_2$. A somewhat different Raman spectrum and assignment has been given in another experiment [152].

The absorption threshold in $PbCl_2$ occurs at ca. 4.5 eV and rises to a sharp peak, observed in the reflection and transmission spectra at room temperature and below [153–157]; the absorption spectrum is affected by moisture, non-stoichiometric composition and crystallization temperature [158–161]. Figure 22 shows the spectral variation of ε_2 of $PbCl_2$ (78 K) as derived [157] from a Kramers-Kronig analysis of the thin film absorption spectrum. The strong narrow exciton peak E_1 at 4.68 eV is seen to be split in the reflection spectrum at 5 K [156], Figure 23, the splitting (0.040±0.005 eV) being attributed to the splitting (into 3 components) of the $^3P_1$ level of the Pb^{2+} ion in the crystalline field of D_{2h} symmetry. The $\varepsilon_2(\omega)$ spectrum at 5 K, Figure 23, also shows a minor peak at 4.820 eV [156] identified as the $n = 2$ member of the (assumed hydrogenic) exciton series giving an energy gap $E_g = 4.863 ± 0.005$ eV compared with a room temperature value of 4.88 eV [162].

On this cationic exciton model the long wavelength $n = 1$ exciton is related to the transition $6\,^1S_0 \rightarrow 6\,^3P_1$ of the lead ion, which is isoelectronic with the mercury atom. The spin-orbit (s.o) splitting of the free lead ion is 3.8 eV but this

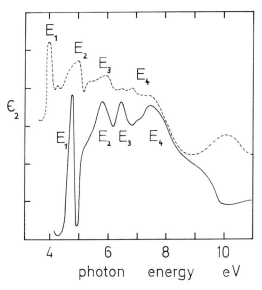

Fig. 22. The spectral variation of ε_2, the imaginary part of the dielectric constant, for PbCl$_2$ (full line) and PbBr$_2$ (dotted line) as derived from the thin film absorption spectra at 78 K [157].

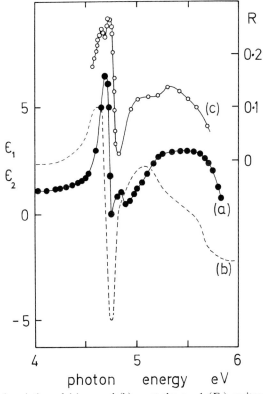

Fig. 23. The spectral variation of (a) ε_2 and (b) ε_1 at the $n = 1$ (E_1) exciton band in PbCl$_2$ (5 K). Curve (c) shows the splitting of the $n = 1$ reflection peak (5 K) [156].

reduces to the mercury atom value of 1.8 eV when the lead ion is introduced into an alkali halide crystal [163]. For this reason peaks E_1, E_3, Figure 22, separation ca. 1.8 eV, are identified with the s.o split levels of the Pb^{2+} ion i.e. E_3 corresponds to $6\,^1S_0 \rightarrow 6\,^1P_1$, transitions.

The complex band E_4 in the $PbCl_2$ spectrum, Figure 22, like that in $PbBr_2$ occurs close in energy to the excitonic halide doublet in KCl and is accordingly identified as such although transitions to the higher energy leve's in lead, such as $6\,^1S_0 \rightarrow 7\,^1P_1$, occur at similar energies.

Lead Bromide

The $\varepsilon_2(\omega)$ spectrum of $PbBr_2$ (78 K) is shown in Figure 22 and the peaks are identified in a similar manner to those in $PbCl_2$. Thus E_1 at 3.98 eV is assigned to an $n=1$ exciton due to the $6\,^1S_0 \rightarrow 6\,^3P_1$ transition in the Pb^{2+} ion; the crystal field splitting of this line is 0.030 eV and the $n=1$ exciton radius is estimated as ca. 7 Å [164], c.f. 8 Å in $PbCl_2$, with $E_g = 4.23$ eV. In very thin polycrystalline $PbBr_2$ films of thickness $d<40$ Å peak E_1 shows a confinement weakening and shift to higher energies, the shift being 0.09 eV for $d \simeq 25$ Å [165]. As in $PbCl_2$ peaks E_1 and E_3 are separated by ca. 1.8 eV so the E_3 band is attributed to the $6\,^1S_0 \rightarrow 6\,^1P_1$ transition in Pb^{2+}. The spacing of the individual maxima at 6.30 and 6.73 eV is close to the spin-orbit splitting value of 0.47 eV in the free Br ion so these two bands are attributed to an anion exciton.

Lead Iodide

PbI_2 has the $Cd(OH)_2$ crystal structure. A large number of structural polytypes occur [166]; crystals grown by the Bridgeman technique, from gels or fairly rapidly from solution are commonly of the 2H polytype (see CdI_2) but larger thin single crystals grown slowly from solution are usually of a higher polytype, particularly if impurities are added to enhance lateral growth. The different optical properties of each polytype are partly responsible for apparently conflicting results [134].

The Raman spectrum of PbI_2 [124] consists of a strong line at 94.3 cm^{-1} (identified as an E_g mode), a broad line at 101 cm^{-1} (identified as A_{1g}) and another line at 76 cm^{-1} attributed to the effect of strain on the E_g mode.

Figure 24 shows the reflectivity spectrum of a PbI_2 (77°K) cleavage plane, perpendicular to the crystal c axis, over the range 2 to 10 eV [167]. The spectrum consists of broad interband maxima, which change little with temperature, and a number of temperature sensitive (excitonic) lines. One such line at 2.495 eV (4965 Å) forms the absorption edge. In the **E** ∥ **c** spectrum of a natural crystal surface [453] exciton peaks at 3.3 and 4.5 eV disappear and the other exciton line intensities alter. This is particularly true of the absorption edge exciton, inset to Figure 24, where the oscillator strength for **E** ∥ **c** is ca. $\frac{1}{4}$ that for **E** ⊥ **c** [168].

Absorption and reflectivity measurements on PbI_2 have located the strong absorption edge exciton peak at positions in the range 4900–4970 Å [169]. It

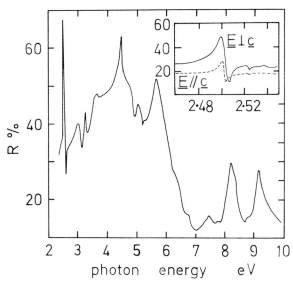

Fig. 24. The reflectivity spectrum of single crystal PbI_2 (77 K). Full line curve – cleaved crystal surface, $E \perp c$ [167]. Inset shows the $E \perp c$ and $E \| c$ band–edge reflectivity spectra of the natural faces of a 2H PbI_2 crystal (4.29 K) [168].

seems that this wavelength variation can be attributed to the effects of sample strain, ageing and different structural polytypes [170]. Thus in the 2H polytype the reflectivity spectrum of a freshly cleaved surface in vacuum yields, via Kramers-Kronig analysis, the absorption spectrum shown in Figure 25 where the $n = 1$ (4966 Å), $n = 2$ (4920 Å), $n = 3$ (4886 Å) exciton lines can be detected. After exposure to air for a few hours the $n = 1$ reflection peak moves to lower energy, by up to 0.010 eV. In another polytype (probably 4H) the $n = 1$ reflection peak occurred at 4942 Å, i.e. close to the thin film position of $n = 1$ marked on Figure 25.

The energies of the $n = 2, 3$ exciton absorption lines, Figure 25, are given by $E_n = 2.552 - 0.127/n^2$ eV [170] but the $n = 1$ line (oscillator strength 1.8×10^{-2} mole^{-1} [171]) occurs 72 meV to the *high* energy side of its hydrogenic position, (64), instead of to the more usual low energy side, (45). One explanation for this is that the lines belong to two overlapping Wannier series (associated with a split upper v.b) with identical binding energies and a separation of 24 meV, the higher energy series being allowed only for $E \| c$ but observed in the $E \perp c$ spectrum because of experimental limitations [173]. However the $E \| c$ spectrum of Figure 24 does not show the expected disappearance of the $n = 1$ peak and enhancement of the $n = 2$ peak which, on this scheme, is the first member of the high energy series. Also reflectivity measurements on $Pb_{1-x}Cd_xI_2$ alloys [168] show that the energy separation of the $n = 1, 2$ lines is constant for $0 < x < 0.15$ indicating that they are both associated with the same v.b and c.b; the $n = 2$ line disappears for $x > 0.15$ (due to compositional disorder scattering) confirming that

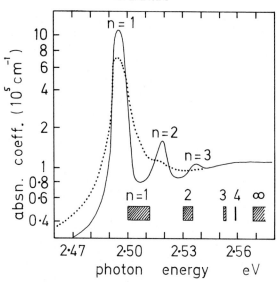

Fig. 25. The single crystal absorption spectrum ($\mathbf{E} \perp \mathbf{c}$) of the band edge excitons in 2H—PbI$_2$ at 4.5 K (solid line) and 77 K (dashed line) [170]. Also marked are the positions of the exciton lines ($n = 1, 2, 3$) in the transmission spectrum of a PbI$_2$ film (thickness 0.16 μm) at 4.2 K. [171] and the reported [172] position of the $n = 4$ line (at 4845 Å) in an evaporated film. The line separations are the same for both series but there is a shift of ca. 20 Å between them.

it has a larger radius than the $n = 1$ exciton which persists to $x = 0.999$, where it is associated with Pb^{2+} impurity. This proves that the $n = 1$ exciton is predominantly 'cationic', consisting of a s-hole and p electron. The requirement of electron-hole orthogonality becomes very stringent when both originate from the same (Pb) atom giving rise to a repulsive central cell correction [174] which is capable of explaining the observed shift to high energy of the $n = 1$ line [168] but would also affect the $n = 2, 3$ lines.

Electro reflectance [170] and electro absorption [177, 178] traces of the $n = 1$ line confirm its excitonic character [179] but do not give consistent values for the binding energy. The pressure coefficient $(\partial E/\partial p)_T$ of the $n = 1$ peak is $-16.5 \pm 0.5 \times 10^{-6}$ eV bar^{-1} at 80 K [180, 181] the negative coefficient being additional evidence for a significant metal 6s state contribution to the upper v.b since states of s-like symmetry rise in energy much faster under pressure than do p or d-like states.

A closer examination of the $n = 1$ exciton line spectra for $\mathbf{E} \parallel, \perp \mathbf{c}$, Figure 24 (inset), shows [175] that the position of the peak in $(\varepsilon_2)_\parallel$ is shifted to higher energy with respect to the peak in $(\varepsilon_2)_\perp$, indicating that the $n = 1$ line contains two excitons, one of symmetry Γ_{3u} at 2.498 eV and another of symmetry Γ_{2u} at 2.501 eV. Correspondingly there are two peaks in $-\text{Im } \hat{\varepsilon}_\perp^{-1}$ and $\text{Im } \hat{\varepsilon}_\parallel^{-1}$ which indicate the longitudinal excitons corresponding to Γ_{3u} and Γ_{2u}. The longitudinal (L)–transverse (T) splittings are 6.2 meV for Γ_{3u} and 1.5 meV for T_{2u} which when substituted in the relation [176].

$$\Delta\omega_{L-T} = f(\omega_p^0)^2/2\varepsilon_0\omega_T \qquad (162)$$

gives oscillator strengths $f_\perp = 1.74 \times 10^{-2}$, $f_\parallel = 4.5 \times 10^{-3}$ for $\varepsilon_0 = 6.25$ and ω_p^0 (the ideal plasma frequency per valence electron) = 3.33 eV.

Above the absorption edge the PbI$_2$ reflectivity spectrum, Figure 24, shows a number of temperature sensitive, excitonic, lines; in 2H—PbI$_2$ (4.5°K) these lines occur at 3.31, 3.96 and 4.48 eV [170] for $\mathbf{E} \perp \mathbf{c}$. A schematic representation of the Γ point transitions giving rise to the 2.5, 3.1 and 3.9 eV excitons [175] is given in Figure 26. A perturbation calculation using only the Pb s and p functions and the energy assignments of Figure 26 correctly predicts the 4:1 ratio of the edge excitons oscillator strengths with $\Delta = 0.77$ eV and a spin-orbit energy parameter $\delta = -0.97$.

Electron energy band calculations [182, 183] put the minimum energy gap at the centre of the hexagonal BZ face at point A (which is group theoretically equivalent to point Γ so Figure 26 still applies) the bands being almost flat along $A-\Gamma$. Structure in the reflectivity spectrum, Figure 24, between 2.5 and 4.5 eV is attributed to transitions between the upper v.b and the three lowest c.b's, as in Figure 26, while structure between 4.5 and 7 eV is assigned to transitions between the next highest (iodine p-like) v.b and the p-like Pb c.b's.

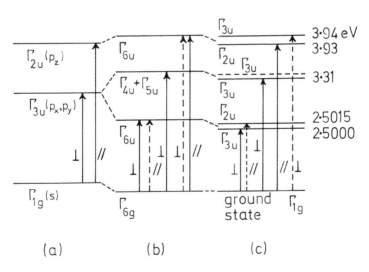

Fig. 26. Schematic representation of the valence and conduction band structure in 2H—PbI$_2$ [175] (a) shows the Γ_{1g} v.b and $^2\Gamma_{3u}$, $^1\Gamma_{2u}$ c.b's composed of s, of p_x, p_y and of p_z functions of Pb respectively; the splitting Δ among the p functions is due to anisotropy (b) shows the splitting resulting from spin-orbit interaction (c) shows the $n=1$ exciton states (and experimental energies) associated with these bands, symmetry requiring that each exciton state is split. Strongly and weakly allowed transitions are denoted by continuous and broken lines respectively.

10.3. Group V halides AsI$_3$, SbI$_3$, BiI$_3$

The crystal structure of these compounds is built up from the stacking of sandwich units each of which consists of two layers of approximately hexagonal close packed iodine atoms with an intervening layer of metal atoms. In AsI$_3$, SbI$_3$ [184, 185] each metal atom is displaced from the centre of the iodine octahedra; in SbI$_3$ [185] the three close iodine atoms give an I—Sb—I angle of 95.8° compared with a molecular bond angle of $99 \pm 1°$ in the vapour phase [186]. A factor group analysis made on the basis of a bimolecular Bravais cell, space group $R\bar{3}$ (point group S_6 which has a centre of symmetry) predicts [188] that the representation of the internal modes reduces to

$$\Gamma(S_6) = 2A_g + 2A_u + 2E_g + 2E_u$$

i.e. each of the four modes characteristic of the molecule is split by crystal symmetry to give non-coincident infrared and Raman frequencies. The measured infrared absorption lines of the three iodides dispersed in Nujol [189] and the Raman lines of single crystal AsI$_3$ [188, 190], SbI$_3$ [191] and powdered BiI$_3$ [190] are given in Table IV together with the line assignments [190].

TABLE IV

The measured infra-red and Raman line frequencies (cm^{-1}) in AsI$_3$, SbI$_3$ and BiI$_3$

Line assignment	AsI$_3$			SbI$_3$		BiI$_3$	
	ν(I.R)	ν (Raman) [190][188]		ν(I.R)	ν (Raman) [190][191]	ν (I.R)	ν (Raman) [190]
ν_{p1} (E_g)		33.3			33		25
ν_{p2} (A_g)		39.0			38		36
ν_6 (E_g)		56.0	50		45.5 43		54
ν_{p3} (E_g)		73.9					
ν_8 (E_u)	74			71		71	
ν_2 (A_g)		84.6	75		67		65
ν_{p4} (A_g)					73		
ν_4 (A_u)	102			89		90	
ν_1 (A_g)		187.1	185		161.5 160		140
ν_3 (A_u)	201			177			
ν_5 (E_g)		108.2	205		139 138		116
ν_7 (E_u)	216(226)			147		115	
$2\nu_2$					130		
$2\nu_3$					352		145
$2\nu_4$ (A_u)							176
$2\nu_4 + \nu_8$						252	
$\nu_1 + \nu_{p2}$					197		
$\nu_2 + \nu_1$					229		

AsI$_3$, SbI$_3$ and BiI$_3$ crystals all cleave (BiI$_3$ the more readily) parallel to (00.1), between adjoining iodine planes, to give thin crystals having their optic, **c**, axis normal to the major surface. Normal incidence transmission and reflection spectra are for $\mathbf{E} \perp \mathbf{c}$; convergent light measurements have shown that SbI$_3$, BiI$_3$ are uniaxial negative, the birefringence $\Delta(n_\perp - n_\parallel)$ being 0.46 ($\lambda = 0.7\ \mu$m) for SbI$_3$ (295 K) [196] and 0.22 ($\lambda = 1.2\ \mu$m) for BiI$_3$ (295 K) [195]. SbI$_3$ is an unstable solid which sublimes, and possibly decomposes, under vacuum while BiI$_3$ is slowly hydrolized in air.

The absorption threshold in AsI$_3$ is characteristic of allowed indirect transitions across an energy gap $E'_g = 2.082$ eV (293 K), the value of E'_g increasing to 2.313 eV at 90 K [192]. At higher energies the absorption edge assumes an exponential shape before reaching a peak at 3.068 eV (90 K).

In SbI$_3$ the exponential absorption threshold ($10^2 < \alpha < 4 \times 10^4$ cm^{-1}) in single crystals [196] rises to a peak A_1, Figure 27, which, in a polycrystalline evaporated SbI$_3$ film (77 K) occurs at 2.572 eV [193]. Absorption peak A_1 is followed by another peak B_1 at 2.938 eV (77 K); both A_1 and B_1 broaden and shift to lower energies with increase in temperature and are identified [196, 193] with $n = 1$

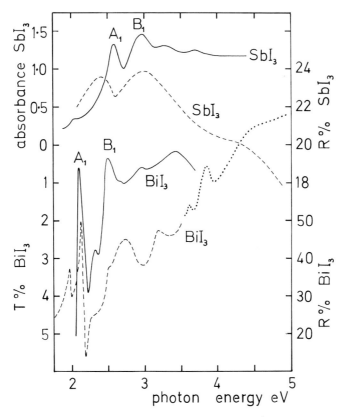

Fig. 27. (a) The absorption spectrum of a SbI$_3$ (77 K) film (200–600 Å thick) evaporated onto a silica substrate (full line) [193] and the reflection spectrum ($\mathbf{E} \perp \mathbf{c}$) of a cleaved SbI$_3$ crystal (293 K) [194] (dashed line). (b) The transmission spectrum ($\mathbf{E} \perp \mathbf{c}$) of a cleaved BiI$_3$ crystal (77 K) [195], thickness 0.35 μm (full line) and the absorption spectrum of a thin BiI$_3$ film (77 K) on a silica substrate [193] (dotted). The reflection spectrum ($\mathbf{E} \perp \mathbf{c}$) of a cleaved BiI$_3$ crystal (77 K) [195] is shown by the dashed line. (Reproduced by permission of The Royal Society)

exciton states associated with the spin-orbit split v.b formed from the 5p orbitals of the iodine ions. The energy separation (0.37 eV) of peaks A_1 and B_1, which are also observed [196, 194] in the single crystal reflection spectrum, Figure 27, gives a measure of the v.b splitting.

The transmission and reflection spectra of single crystal BiI_3 (77 K) [195] over the range 1.9 to 4 eV is shown in Figure 27 together with the transmission spectrum ($3.6 \leq \hbar\omega \leq 5$ eV) of a BiI_3 film deposited on a heated silica substrate [193]. The peak absorption coefficients of A_1, B_1 are 1.8×10^5 cm^{-1}. The absorption edge formed by the low energy side of A_1 varies exponentially with $\hbar\omega$ for $10 < \alpha < 10^4$ cm^{-1} [195]; at small values of α the step-like nature of the absorption threshold has been ascribed to indirect exciton and interband transitions [197].

On the low energy side of the absorption threshold a number of absorption lines have been observed some of which [198] formed an 'inverted' hydrogenic series described by

$$\nu_n = \nu_\infty + \frac{R}{n^2} = \left(15\,978 + \frac{1995}{n^2}\right) \text{cm}^{-1}$$

with $n = 3, 4, 5, 6, 7$; these lines show an unequal shift to lower energies with applied uniaxial stress along **c** [199]. The lines have been explained [198] on a bielectron (or bihole) model in which two electrons whose curvatures are such as to give a negative reduced mass are coupled to each other by Coulomb interaction. However the strong variation of absorption line intensity with sample preparation [200, 201, 202] together with photocurrent [201] and luminescence measurements [202] suggest that the lines may be associated with impurities such as excess Bi [200].

Absorption bands A_1 (2.095 eV), B_1 (2.479 eV) in BiI_3 (77 K) are identified [195] with $n = 1$ exciton states of the spin-orbit (s.o) split v.b; the large negative pressure coefficient of A_1 [180] indicating (c.f PbI_2) a significant contribution from metal $6s$ states to the upper v.b. The energy separation of A_1 and B_1 (0.38 eV), which gives a measure of the s.o splitting, is close to that in SbI_3. Assuming that absorption bands A_1 (2.095 eV), A_2 (2.281) and B_1 (2.479 eV), B_2 (2.653) form the first two members of a hydrogenic series, (52b), gives the comparatively large Rydberg constants $R_e^A = 2 \times 10^3$ cm^{-1}, $R_e^B = 1.86 \times 10^3$ cm^{-1} and $E_g^A = 2.33$ eV, $E_g^B = 2.696$ eV. Substituting these values of R_e and the measured values of $n_\perp$ (= 2.48) and $n_\parallel$ (= 2.26) [195] in (52c) gives $r_\perp^{(1)A} = 5.18$ Å, $r_\parallel^{(1)A} = 4.72$ Å and $r_\perp^{(1)B} = 5.57$ Å, $r_\parallel^{(1)B} = 5.07$ Å. These values of r are comparable with the mean radius (5.5 Å) of the approximately spherical distribution of next nearest Bi ions surrounding each I ion, which may invalidate the use of (52). Magneto-absorption measurements at the threshold give longitudinal and transverse carrier masses of 0.083 m_0 and 0.014 m_0 respectively [203].

The halogen doublet A_1, B_1 which is such a distinctive feature of the SbI_3, BiI_3 spectra has also been observed [193] in the absorption spectra of the other layer crystals CaI_2, CdI_2, FeI_2, MnI_2, PbI_2. The energy at which component A_1 occurs in all these compounds, including SbI_3, BiI_3 is roughly proportional to the inverse square of the interionic distance in the crystal [193].

In the absence of any band structure calculations for SbI_3, BiI_3, the absorption and reflection peaks which occur to the high energy side of B_1 remain unidentified.

10.4. TRANSITION METAL HALIDES

The dichlorides, dibromides and di iodides of some transition metals occur as layer structures [119]. The optical properties of some of these have been studied viz. $MnCl_2$, $FeCl_2$, $CoCl_2$, $NiCl_2$, $NiBr_2$. NiI_2 which have the $CdCl_2$ crystal structure and $MnBr_2$, MnI_2, FeI_2, $CoBr_2$ which have the $Cd(OH)_2$ crystal structure.

TABLE V

Experimental values (eV) of the transitions to excited levels in Mn^{++} in $MnCl_2$, $MnBr_2$ at 78 K [204, 205]

Excited levels	$MnCl_2$	$MnBr_2$
$^4\Gamma_{1g}\,(^4G)$	2.293	2.287
$^4\Gamma_{2g}\,(^4G)$	2.727	2.683
$^4A_{1g}\,(^4G)^*$	2.924	2.861
$^4_aE_g\,(^4G)^*$	2.953	2.919
$^4\Gamma_{2g}\,(^4D)$	3.316	3.287
$^4_bE_g\,(^4D)^*$	3.479	3.409
$^4\Gamma_{1g}\,(^4P)$	4.524	3.712
$^4A_{2g}\,(^4F)^*$	4.760	4.313
$^4\Gamma_{1g}\,(^4F)$	5.038	4.636
$^4\Gamma_{2g}\,(^4F)$	5.252	4.803

On the low energy side of the fundamental absorption edge a number of weak absorption lines are observed at low temperatures which are characteristic of the metal ion. In $MnCl_2$, $MnBr_2$ for example these lines occur at the energies given in Table V. The crystal field, of trigonal symmetry, produces a splitting of some of the lines (particularly of the $^4A_{1g}$ and ^4E_g levels) but if this is ignored the observed line energies can be derived from the free ion values [206] by using the covalency ε [211] and cubic ligand field Dq as adjustable parameters, the starred levels in Table V depending only on ε [212, 213]. The observed pressure shift of the metal ion lines in $MnCl_2$, $MnBr_2$; $CoCl_2$, $CoBr_2$; $NiCl_2$, $NiBr_2$ can be attributed to an increase in crystal field strength and covalency [214]. In the case of $CoCl_2$, $CoBr_2$ single crystals (20 K) four parameter crystal field theory gave a comparatively poor description of the observed cobalt ion absorption lines [215], however measurements on polycrystalline samples [210] confirmed the expected similarity of the $CoCl_2$, $CoBr_2$ line spectra.

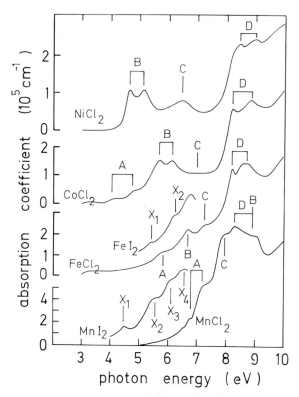

Fig. 28. The absorption edge spectra of evaporated films of the $CdCl_2$ type compounds $NiCl_2$, $CoCl_2$, $FeCl_2$ and $MnCl_2$ at room temperature [208] and of the $Cd(OH)_2$ type compounds MnI_2, FeI_2 at liquid nitrogen temperature [193].

The absorption edge spectra of the $CdCl_2$ type compounds $NiCl_2$, $CoCl_2$, $FeCl_2$, $MnCl_2$ [208] and the $Cd(OH)_2$ type MnI_2, FeI_2 [193] are shown in Figure 28. In MnI_2, FeI_2 absorption peaks labelled $X_1 --- X_4$ are identified as in CdI_2, Figure 18, i.e. with the crystal field split halogen doublet. In the $CdCl_2$ type compounds of Figure 28, which have similar lattice constants, the shifts in peaks B and C through the series ($NiCl_2$ to $MnCl_2$) parallels the shift in cation $3d^n$ and $3d^{n-1}\,4s$ level binding energies; peaks B and C are accordingly assigned to allowed transitions from the Cl^{2-} $3p$ level to the metal $3d$ and $4s$ levels respectively, the doublet nature being due to crystal field splitting [208]. An absorption doublet D which occurs at almost the same energies throughout the series ($NiCl_2$ to $MnCl_2$) has not been positively identified neither has the doublet A which occurs at the absorption threshold.

Figure 29 shows the absorption spectra of the $CdCl_2$ type compounds up to 35 eV [208], 67 eV in the case of $MnCl_2$ [209], together with that of $MnBr_2$ [209]. All the $CdCl_2$ type spectra show a maximum in absorption around 15 eV which, by comparison with the free ion spectra, can be identified with the $3d \rightarrow 4p$ transition in the metal ion. At energies above ca. 15 eV the absorption falls

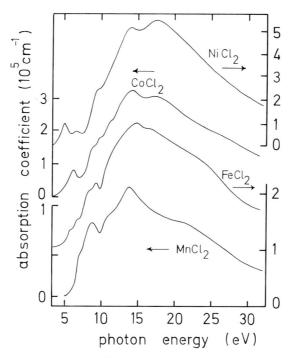

Fig. 29a. The room temperature absorption spectra of $MnCl_2$, $FeCl_2$, $CoCl_2$, $NiCl_2$ evaporated films [208].

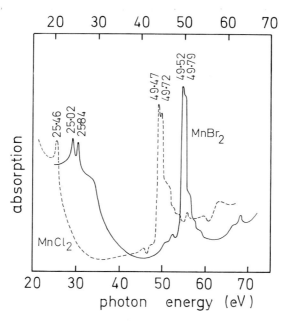

Fig. 29b. The absorption spectra of evaporated films of $MnCl_2$, (dashed line) and $MnBr_2$ (full line) [209]. Note that the $MnBr_2$ (upper) energy scale is displaced from that of $MnCl_2$. The structure around 25 eV is due to second order effects and should be ignored. [454]

slowly to a minimum at around 35 eV in MnCl$_2$ and 45 eV in MnBr$_2$, the MnCl$_2$ spectrum showing an unidentified shoulder at ca 21 eV. The shoulder at 28.9 eV in MnBr$_2$ is attributed to the excitation of the 4s electrons of Br to p states in the conduction band (free ion value 27.3 eV).

From 45 to 56 eV the MnCl$_2$, MnBr$_2$ spectra are similar. In both cases all the structure in this region is attributed [209] to excitation of the Mn^{2+} 3p electrons, either to 3d states forming the bottom of the c.b or, in the case of the sharp (doublet) peak at ca. 50 eV, to associated exciton states.

The Cl$^-$ L_{23} absorption spectra of NiCl$_2$, CoCl$_2$, FeCl$_2$ and MnCl$_2$ are very similar [207], each spectrum consists of a number of sharp bands in the range 200 to 210 eV followed by a broad band at higher energies. The sharp bands are attributed to the creation of excitons with holes in the 2p level of chlorine, the origin of the broad band is uncertain.

10.5. Group III chalcogenides GaS, GaSe, GaTe

The III–VI semiconducting compounds GaS, GaSe GaTe (type MY) all occur as layer-type crystals. In the case of GaS, GaSe each layer consists of four planes of atoms arranged in the sequence Y, M, M, Y; the tetrahedrally coordinated metal M atoms forming the two central planes of each layer. Three different layer stacking arrangements occur [216] usually referred to as the β (or hexagonal GaS, D_{6h}^4), $\varepsilon(D_{3h}^1)$ or $\gamma(C_{3v}^5)$ forms; the stacking of adjacent layers in the rhombohedral γ polytype is the same as that in the hexagonal ε polytype. The bonding between the layers is very weak, which accounts for the easy cleavage of these crystals, but the M—M and M—Y bond lengths have the values predicted from covalent radii [218]. A further difference occurs in GaTe [217] where portions of each *layer* are joined together by regions (strips parallel to the crystal b axis) in which the Ga—Ga bonds are approximately at right angles to the usual Ga—Ga bond directions.

In the β polytype (which has an inversion centre located on the mid plane between layers) where two layers and four molecular units are contained in a primitive unit cell, the 24 normal modes of vibration at the B.Z centre are described by [219]

$$\Gamma = 2A_{1g} + 2A_{2u} + 2B_{1u} + 2B_{2g} + 2E_{1g} + 2E_{1u} + 2E_{2g} + 2E_{2u}$$

there are six non-degenerate Raman active modes ($2A_{1g}$, $2E_{2g}$, $2E_{1g}$) and two infrared active modes (E_{1u}, A_{2u}).

The 24 normal modes of vibration of the ε polytype (which lacks an inversion centre) are described by [220]

$$\Gamma = 4A_1' + 4A_2'' + 4E' + 4E''$$

in this case there are 11 non-degenerate Raman active modes, $4A_1'$, $3E'$, $4E''$ and there are 6 non-degenerate infrared active modes $3A_2''$, $3E'$.

In the γ polytype all modes are Raman active [120].

A number of measurements have been made of the infrared absorption and Raman scattering spectra of GaS, GaSe which allow the structural polytype to be identified [120, 219–227]. Apart from some differences in experimental values and assignments the observed frequencies (cm^{-1}) of the (polarization dependent) Raman lines in GaS (293 K) can be assigned to the β polytype [220, 120, 226, 227] viz. E_{2g}^2 (22.0 cm^{-1}), E_{1g}' (75.2), A_{1g}' (187.9), E_{1g}^2 (290.5), E_{2g}' (295.0), A_{1g}^2 (360.7) while in GaSe (77 K) the Raman lines are best described by the ε polytype [225, 226, 227, 120] viz. E' (19.5 cm^{-1}), E'' (60.1), A_1' (134.3), E'' (211.9), E' (TO, 215), E' (LO, 252.1), A_1' (308.0). This interpretation of the GaS, GaSe Raman spectra is supported by infrared measurements, the ε polytype of GaSe (4 K) shows absorption lines [227] at 20.1 cm^{-1} ($\mathbf{E} \perp \mathbf{c}$), 36.7 cm^{-1} ($\mathbf{E} \| \mathbf{c}$) and 212.1 cm^{-1} ($\mathbf{q} \| \mathbf{c}, \mathbf{E} \perp \mathbf{c}$, 293 K); the 20.1 and 36.7 cm^{-1} lines are attributed to Raman activity of the E' and A_2'' modes whereas the corresponding E_{1g} and B_{2g} modes of GaS are not infrared active.

The fundamental absorption edges in GaS, GaSe are shown in Figure 30. In

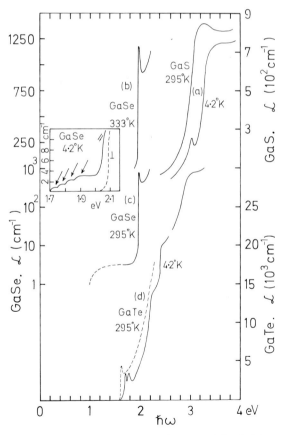

Fig. 30. The fundamental absorption edges in (a) GaS (295, 4.2 K) [229]; (b) and (c) hexagonal GaSe (333 K) [255] and (295 K) [229] with inset showing the absorption threshold (4.2 K) for $\mathbf{E} \| \mathbf{c}$ and $\mathbf{E} \perp \mathbf{c}$ (dashed line) [237]; (d) GaTe at 295 and 4.2 K [229].

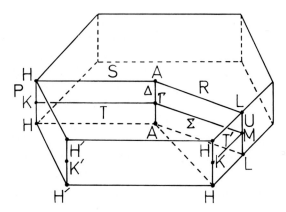

Fig. 31a. The first Brillouin zone of hexagonal GaS and GaSe.

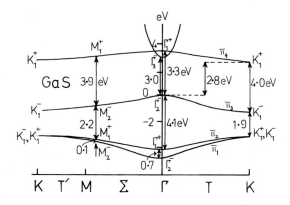

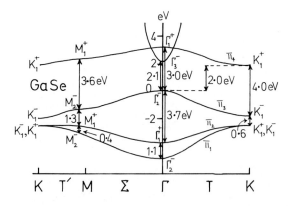

Fig. 31b. The π electron energy band structures of hexagonal GaS and GaSe [234]. The π energy bands are associated with electrons in the p_z orbitals of Ga and S(Se) perpendicular to the crystal layer. The lower lying (filled) σ bands, not shown, are associated with bonding s, p_x, p_y in the layer. In addition to the four π bands the light effective mass c.b is shown schematically. Authors correction:– for $K_1^\pm$ in π_1, π_2 bands read $K_2^\pm$, for K_1^- in π_3 band read K_3^-, for K_1^+ in π_4 band read K_3^+ [455].

GaS the absorption threshold rises to an exciton peak [228, 229] at 3.03 eV (4.2 K), the position of the reflection peak being the same [230] for $\mathbf{E}\|\mathbf{c}$, $\mathbf{E}\perp\mathbf{c}$. Measurements of the absorption threshold in GaS at different temperatures show [229, 231, 232] that the threshold can be described [231] in terms of forbidden indirect transitions across an energy gap $E'_g = 2.65$ eV (4.2 K) [229] $= 2.8$ eV (0 K) [233, 234]. On the calculated band structure of GaS, GaSe [234], Figure 31, the indirect edge corresponds to the transition from the Γ_2^- state of the π_3 band to the K_1^+ state of the π_4 band which is allowed only for $\mathbf{E}\|\mathbf{c}$. Away from the symmetry points indirect $\pi_3 \to \pi_4$ transitions are allowed for $\mathbf{E}\|\mathbf{c}$, $\mathbf{E}\perp\mathbf{c}$. The oscillator strength of any $\pi \to \pi$ transition is zero if the π bands are purely p_z in character and overlap between p_z functions on different sites is neglected. Some mixing of an s-character band into the π bands is necessary to give any intensity to the $\pi - \pi$ transitions which accounts for the 'forbidden' nature of the indirect edge in GaS [234].

Transmission measurements ($\mathbf{E}\|\mathbf{c}$) on Bridgeman single crystals of GaS_xSe_{1-x} (77 K) [235] which occur as mixed ε, γ polytypes for $0 \le x \le 0.1$, mixed ε, γ, β polytypes for $0.2 \le x \le 0.4$ and as the β polytype for $0.6 \le x \le 1$, have followed the linear shift of the indirect edge to higher energies with increasing x. In the mixed ε, γ crystal the threshold is ca. 50 meV to the low energy side of that in the β modification. Similarly the Γ exciton peak ($\mathbf{E}\perp\mathbf{c}$), which also shifts linearly to higher energies (and weakens) with increasing x, occurs at a lower energy (ca. 50 meV) in the mixed ε, γ crystal.

In GaSe, Figure 30, the absorption threshold due to indirect transitions [236, 237] is only observed for $\mathbf{E}\|\mathbf{c}$ as predicted by the selection rules for π_3 (Γ_2^-) $\to$ $\pi_4(K_3^+)$ transitions. The form of the threshold is best described [237, 238] in terms of phonon assisted transitions into two dimensional exciton states and the energy independent density of states appropriate to a single two dimensional exciton band, the absorption coefficient is given by

$$\alpha(\omega) \propto (n+\tfrac{1}{2})^{-3} S[\hbar\omega - E'_g + R(n+\tfrac{1}{2})^2 + \hbar\omega_i] \qquad (n = 0, 1, 2, 3, \ldots),$$

where $S[x]$ is a step function such that $S[x] = 1$ for $x > 0$ and $S[x] = 0$ for $x < 0$ and ω_i is the phonon frequency of the ith K_3^+ mode at the K point. Considering only exciton state $n = 0$ the four (as predicted) phonon emission steps (arrowed) in the threshold (4.2 K), Figure 30, give [237] $R = 0.05$ eV, $E'_g + \hbar\omega_1 = 1.93 \pm 0.01$ eV. The hydrostatic pressure coefficient of the indirect threshold ($\alpha < 200$ cm^{-1}) in GaSe is $(-6.5 \pm 0.6)10^{-6}$ eV bar^{-1} [239] while that of the following Γ exciton peak is $(-4.2 \pm 0.3)10^{-6}$ eV bar^{-1}. With increasing pressure the exciton peak diminishes due to overlap with the indirect transition continuum (as occurs in GaS), the negative pressure coefficients are attributed to the (pressure sensitive) splitting of the c.b and v.b levels by layer–layer interaction [239].

Figure 32 shows the ε_1, ε_2 spectra of GaS (300 K) as derived [241] from the reflectivity spectrum ($\mathbf{E}\perp\mathbf{c}$) [242] over the range 2 to 12 eV. In establishing the

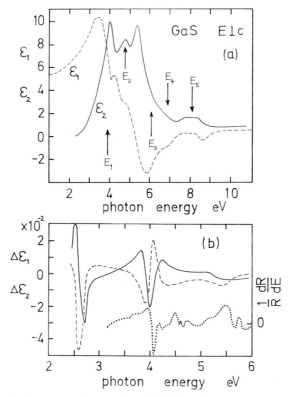

Fig. 32. (a) The spectral variation of ε_1, ε_2 for GaS (293 K) derived from the normal reflectivity $\mathbf{E} \perp \mathbf{c}$ [241]. (b) The wavelength modulated reflection spectrum (dotted) of GaS (5 K), $\mathbf{E} \perp \mathbf{c}$ [243]. The changes $\Delta\varepsilon_1$ (dashed line) and $\Delta\varepsilon_2$ (full line) derived from the thermo-reflectance spectrum of GaS (77 K) [241].

band structure of Figure 31 the peaks in ε_2 at 3.9 eV and 6.1 eV were attributed to $M_2^-(\pi_3) \to M_1^+(\pi_4)$ and $M_1^+(\pi_2) \to M_1^+(\pi_4)$ transitions respectively. The remaining peak identities can be deduced from Figure 31. Also shown in Figure 32 are the temperature induced changes $\Delta\varepsilon_1$, $\Delta\varepsilon_2$ derived from the thermoreflectance spectrum of GaS (77 K) [241]. The dispersion-like shape of $\Delta\varepsilon_2$ around 2.6 eV can be described in terms of a temperature shift (without broadening) of an M_0 exciton line at 2.646 eV. Similarly electron-hole interaction at an M_1 critical point (3.998 eV) is required to give the $\Delta\varepsilon_2$ line shape at 3.9 eV. The wavelength modulated spectrum of GaS (5 K) [243] shows a sharp peak at 4.04 eV which is the counterpart of the thermoreflectance peak but the additional structure has not been identified.

Above the indirect threshold the absorption in GaSe rises to a Γ exciton peak, Figure 30; at low temperatures the higher quantum number members of the exciton series are observed, the n^{-2} energy dependence of these lines being characteristic of three, rather than two, dimensional excitons, (53). This three dimensional character of the excitons in GaSe is confirmed by the fine structure of the exciton ground state which occurs in crystals of mixed $\varepsilon - \gamma$ stacking [245], Figure 33.

A three dimensional calculation (which includes layer–layer interaction) of the electron energy bands in GaSe [247] gives twice the number of bands of the two dimensional calculation, Figure 31. The top of the v.b has symmetry Γ_4^- and its partner, which would join to form a single level in a two dimensional calculation, appears 0.8 eV lower and has symmetry Γ_1^+. Four doubly degenerate bands, Γ_5^+, Γ_6^+ and Γ_5^-, Γ_6^- are located between the two bands. The c.b level Γ_3^+ was set slightly higher in energy than the c.b minimum at M_3^+ (associated with the indirect threshold). Figure 33 shows the form of the Γ exciton reflection spectrum in $\varepsilon - \gamma$ GaSe for different polarizations [246]. With $\mathbf{E} \perp \mathbf{c}$ only the weak lines t_i are observed, these are some of the lines due to transitions into triplet states of the 1s (Γ_6) exciton; the multiplicity of lines is due to the number of possible layer stacking configurations encompassed by the 1s exciton [245], this determines the effective dielectric constant and hence the exciton ground state energy, (53). In principle such components should also occur in the $n = 2, 3$ lines but since these excited states extend over a very much larger number of layers the component separations will be very small. Changing the polarization to give a component of

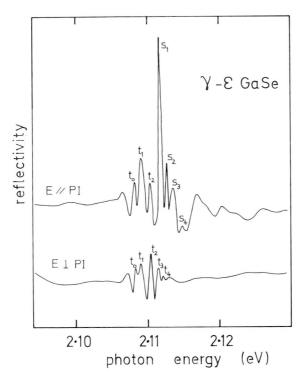

Fig. 33a. Fine structure of the exciton ground state in $\gamma - \varepsilon$ GaSe. Measurements were made with light incident at an angle of 45° with respect to the crystal c axis and the polarization vector parallel ($\mathbf{E} \parallel PI$) and perpendicular ($\mathbf{E} \perp PI$) to the plane of incidence [246].

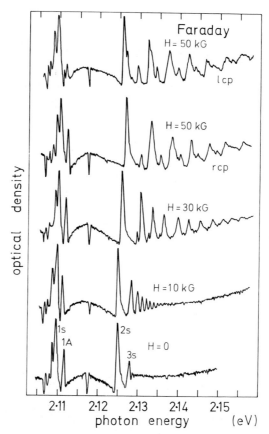

Fig. 33b. The magneto absorption spectrum of $\gamma-\varepsilon$ GaSe (2 K) in the Faraday geometry (light propagation vector $\|\mathbf{H}\|\mathbf{c}$). The uppermost spectrum was obtained with left circularly polarized (l.c.p.) light, all the others with right circularly polarized (r.c.p) light [262].

$\mathbf{E}\|\mathbf{c}$ allows transition into the singlet states Γ_4 i.e. the strong lines marked s_i, Figure 33, becomes visible [257, 258]. Fitting the exciton lines to a hydrogen-like series, (53) gives $E_g = 2.1295$ eV (1.5 K) and the radius of the 1s exciton as 32 Å i.e. the exciton extends over approximately four layers. In a single crystal of the β polytype, where all the layers are equivalent, no fine structure occurs in $n = 1$ but the exciton lines are displaced to higher energy by 50 meV [245], the exciton ionization energy is the same as in the $\gamma - \varepsilon$ modification.

A number of measurements have been made of the electroreflectance (er) and electroabsorbance (ea) in the vicinity of the GaSe band edge [248–253]; the er line shapes are analysed to give E_g and the exciton binding energy, for example 2.109 eV (77 K) and 0.021 eV respectively [249]. An analysis of the thermoreflectance spectrum in the Γ exciton region of GaSe (77 K) gives the $\Delta\varepsilon_1$, $\Delta\varepsilon_2$ spectra shown in Figure 33, [254]. The $\Delta\varepsilon_2$ line shape is described by an energy shift (without broadening) of the ground state exciton line giving $E_g = 2.093$ eV and γ (the Lorentzian line width parameter) $= 41$ meV. Measurements at different

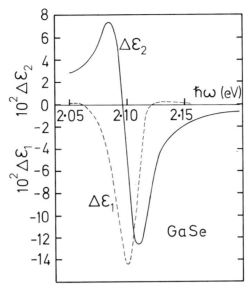

Fig. 33c. The changes $\Delta\varepsilon_1$ (dashed line) and $\Delta\varepsilon_2$ (full line) in the real and imaginary parts of the dielectric constant of GaSe (77 K) derived from the thermoreflectance spectrum, $\mathbf{E}\perp\mathbf{c}$ [254].

temperatures give $dE_g/dT = -5.49 \times 10^{-4}$ eV deg^{-1} compared with -4.14×10^{-4} eV deg^{-1} obtained from the temperature shift of the exciton absorption peak [255]. Thermoreflectance measurements from the face of a GaSe crystal normal to the layers [256] confirm the results of Figure 33 viz. the $\Delta R/R$ structure for $\mathbf{E}\|\mathbf{c}$ is ca. 20 times greater than for the other polarization.

Magneto absorption and Faraday rotation measurements have been made in the vicinity of the GaSe band edge excitons [244, 259–263]. Figure 33 shows the magneto absorption, $\mathbf{E}\perp\mathbf{c}$ (Faraday configuration) of $\varepsilon - \gamma$ GaSe (2 K) [262], the line marked 1A is the $n = 1$ state of the longitudinal exciton which occurs because of sample misalignment. The shifts between the corresponding r.c.p. and l.c.p. lines of the 1s exciton states (which have no orbital moment) is entirely due to the spin term which is non-zero for the triplet states whose spin components S_z are ±1; this r.c.p.-l.c.p. splitting gives the z component of the exciton g-factor as 2.7±0.2 [246]. The field dependence of the 1s, 2s, 3s exciton lines is shown by the experimental points of Figure 34. The singlet 1s state is recognized by the absence of spin-splitting, it shows only the diamagnetic shift (131). In the effective mass approximation the diamagnetic shift is σH_z^2 for 1s excitons and $14\sigma H_z^2$ for 2s excitons where $\sigma = \hbar^4 \varepsilon_x \varepsilon_z / 4\mu_x^3 e^2 c^2$. The curves through the experimental points of Figure 34 were calculated with $\sigma = 4.3 \times 10^{-5}$ meV (kG)^2 which gives $\mu_x = 0.13\, m_0$ for $\varepsilon_x = 10.2$, $\varepsilon_z = 7.6$ [264]; this is similar to the value (0.14 m_0) obtained [265] from the zero field exciton Rydberg constant.

Figure 35 shows the spectral variation of ε_1, ε_2 for GaSe (293 K) derived from the reflectivity over the range 2 to 80 eV [242, 266]. The most detailed reflectivity

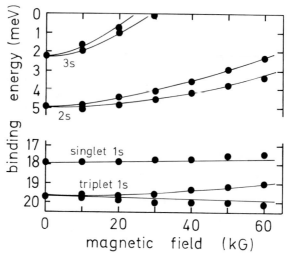

Fig. 34. The calculated magnetic field dependence of exciton states $n = 1, 2, 3$ in GaSe (full lines) fitted to the experimental values (solid circles) [262] – see Figure 33, at $H = 0$ [246].

measurements [267] below 12 eV have located structure at (eV): $-E'_1$ (3.3), E_1 (3.66), E_2 (4), E'_2 (4.56), E_4 (5.3), E_5 (6.7), E_6 (8.3), E_7 (8.9), E_8 (10.8), the peak intensities, but not energies, being affected by surface conditions. At higher energies structure occurs at the values marked in Figure 35. Peaks E_{10}, E_{11}, E_{12} are evidently due to the excitation of Ga $3d$ electrons (seen also in GaP, GaAs, GaSb [268–271] while band E_{13} can be attributed to the excitation of Se $3d$ electrons into the c.b. On the two dimensional band structure of GaSe, Figure 31, peaks E_1 and E_3 were attributed, as in GaS, to $M_2^-(\pi_3) \to M_1^+(\pi_4)$ and $M_1^+(\pi_2) \to M_1^+(\pi_4)$ respectively, the remaining peak identities can be taken from Figure 31. A speculative attempt to assign reflection peaks E_1-E_9 to transitions on a three dimensional band structure [247] has also been made [266]. Other reflectivity measurements have shown [273] that for GaSe peak E_1 is strong for $\mathbf{E} \perp \mathbf{c}$, weak for $\mathbf{E} \| \mathbf{c}$ whereas E_3 is the same for both polarizations, as is E_1 for GaS. The peak in the derived energy loss function $-\operatorname{Im} \hat{\varepsilon}^{-1}$, Figure 35, at 16 eV is close to the observed electron energy loss maximum at 15.8 eV [272].

Single crystal transmission measurements in the vicinity of E_1 in GaSe [274, 275] have detected a shoulder at ca. 3.2 eV (292 K), c.f E'_1, which sharpens to a peak (labelled E_h, inset to Figure 35) at 3.378 eV (77 K) i.e. 0.318 eV below E_1. The peak E_h has been identified as a hyperbolic exciton state of the M_1 type (E_1) transition and this is supported by the thermoreflectance [241] and wavelength modulated [243], Figure 35, E_h line shapes.

Another interpretation based on the similarity between the 2.0 eV (M_0 exciton) and 3.38 eV line shapes allows the possibility of E_h being associated with an M_0 exciton [275]. The electroabsorption line shape has also identified E_h with a hyperbolic exciton [276] whereas the electroreflectance (er) line shapes have attributed E_h, E_1 to spin-orbit splitting of the valence band [253]. In the er

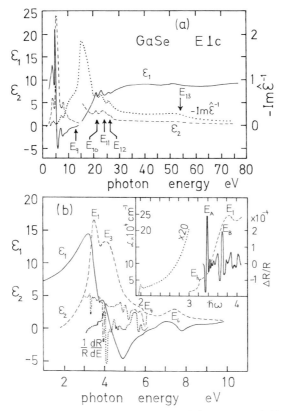

Fig. 35. (a) The spectral variation of ε_1, ε_2 and $-\operatorname{Im}\hat{\varepsilon}^{-1}$ for GaSe (293 K) $\mathbf{E}\perp\mathbf{c}$ [266]. For $\hbar\omega > 15$ eV the values of ε_1, ε_2 have been multiplied by ten for clarity. (b) The spectral variation of ε_1, ε_2 for GaSe (293 K) over the range 1 to 11 eV [241] together with the wavelength modulated reflectivity of GaSe (5 K) and, for comparison that of GaS (5 K) – dotted curve [243]. Inset shows the absorption edge, $\mathbf{E}\perp\mathbf{c}$ in GaSe (295 K) [275] together with the electroreflectance spectrum of GaSe, $\mathbf{E}\perp\mathbf{c}$ at 90 K [277].

spectrum of Figure 35 [277] the optical constants of GaSe are such that peak E_A (3.36 eV at 90 K) is dominated by $\Delta\varepsilon_1$, resembles $-F(-x)$, (98), and is therefore identified with an $M_{1\perp}$ c.p, Figure 11. If the optical constants are such that er line E_B (3.7 eV) is also dominated by $\Delta\varepsilon_1$ then its line shape is a composite of M_2 and M_3 c.p's, in which case it is not the s.o split component of E_A.

At energies above E_1 the wavelength modulated reflectivity spectrum, Figure 35, shows peaks at 4.88, 5.15, 5.56 and 5.80 eV which on the two dimensional band structure of Figure 31 can be related to the transitions $M_1^+(\pi_2) \to M_1^+(\pi_4)$; K_1^+, $K_1^-(\pi_2) \to K_1^+(\pi_4)$ and $M_2^-(\pi_1) \to M_1^+(\pi_4)$.

GaTe

The fundamental absorption edge in cleaved GaTe crystals (4.2 K) for unpolarized light incident normal to the crystal layers [278] is given in Figure 30. The peak occurring at 1.663 eV (300 K) [279–282] splits on cooling to low temperatures

[279, 282], the components occurring at 1.779 and 1.795 eV at 4.2 K [282]; the line strengths are influenced by mechanically working the crystal and by annealing the crystal in an atmosphere of hydrogen [282]. Identifying the component lines with $n = 1, 2$ three dimensional exciton states [283] gives $E_g = 1.797$ eV (77 K) compared with a value of 1.775 eV (4.2 K) estimated from the absorption threshold. The thermoreflectance spectrum ($\mathbf{E} \perp \mathbf{c}$) of GaTe (284 K) [284] at the band edge has been interpreted in terms of an energy shift, without broadening, of a Lorentzian shaped ground state exciton line at 1.676 eV. The photoreflectance spectrum (which is thought to resemble the electroreflectance spectrum) of GaTe (295 K) around this energy [285] shows a positive peak at 1.679 eV which may be due to the $n = 1$ exciton line.

The reflectivity of a cleaved GaTe crystal (293 K) has been measured over the range 1 to 6 eV [286, 287] to give the ε_1, ε_2 spectra shown in Figure 36. Also shown in Figure 36 is the wavelength modulated reflectivity spectrum of GaTe at room temperature [288]. Both spectra show sharp structure due to the exciton at

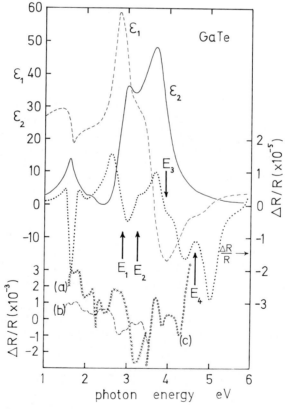

Fig. 36. Dashed and full line curves give ε_1, ε_2 for GaTe ($\mathbf{E} \perp \mathbf{c}$) at room temperature [287]. Dotted curve gives the wavelength modulated normal incidence reflectivity of a cleaved GaTe crystal at room temperature [288]. Curves (a), (b), (c) represent the thermoreflectance spectra of GaTe at room temperature for $\mathbf{E} \perp \mathbf{b}$, $\mathbf{E} \| \mathbf{b}$ and unpolarized radiation respectively [289].

1.680 eV with further structure at 2.90 (E_1), 3.34 (E_2) 3.90 (E_3) and 4.70 eV (E_4). The $\Delta\varepsilon_1$, $\Delta\varepsilon_2$ line shapes derived from the $\Delta R/R$ spectrum indicate that E_1 is due to a metamorphism of two-dimensional M_0 and M_1 c.p's at 2.922 eV while E_3 is associated with a two dimensional M_2 c.p at 3.873 eV.

Figure 36 shows the thermoreflectance of GaTe (room temperature) for light polarized parallel and perpendicular to **b** – the crystal axis contained in the layer plane [289]. Strong polarization effects occur, some of the structure corresponds to that in the absorption, Figure 30, and reflection, Figure 36, spectra but the absence of a calculated band structure makes the assignment of structure in Figure 36 uncertain.

10.6. GROUP IV DICHALCOGENIDES

SnS_2, $SnSe_2$ are members of the class of Group IV dichalcogenides $M^{iv}X_2^{vi}$ (M = Ti, Zr, Hf, Si, Sn; X = S, Se, Te, although not all combinations occur) having the CdI_2, type C6 structure. This is composed of planes of hexagonal close packed X atoms interleaved with planes of metal M atoms to give the stacking sequence X—M—X, X—M—X in the **c** direction of the unit cell; the axial ratio c/a is not very different from the value 1.63 expected for the close packing of spherical X atoms. The C6 structure, with its adjoining planes of weakly bound X atoms, which allows easy cleavage in {00.1}, is not readily explained in terms of an ionic model and it is significant [290] that MX_2 compounds having this structure are those for which the 6-fold octahedral coordination of M is characteristic. It is concluded therefore that the C6 structure results from the M—X bonding.

Single crystals of n-type SnS_2, $SnSe_2$ can be grown by iodine vapour transport [291]. These flexible crystals are easily cleaved, using transparent adhesive tape, to give thin samples having excellent (00.1) reflecting surfaces. The room temperature reflection spectra ($\mathbf{E}\perp\mathbf{c}$) of SnS_2 [292] and $SnSe_2$ [293] are given in Figure 37. Transmission measurements ($\mathbf{E}\perp\mathbf{c}$) on thin single crystals have given $\alpha(\omega)$, for $\hbar\omega < 5$ eV, in SnS_2 [292, 294, 295] and $SnSe_2$ [294–297].

In the case of SnS_2 the absorption edge has been analysed in terms of allowed [292, 295] (forbidden [294]) indirect transitions across an energy gap E'_g (300 K) = 2.21 [292], 2.22 [295], 2.07 eV [294] followed by forbidden direct transitions across an energy gap $E_g = 2.88$ eV [294]. The ordinary refractive index $n_\perp$ of SnS_2 (300 K) computed [294] from reflectivity data increases from 2.72 at $\lambda = 2.1$ μm to 3.15 at $\lambda = 0.69$ μm at which wavelength the birefringence Δ ($= n_\perp - n$) = 0.69 [295].

Measurements on the single crystal alloy SnS_xSe_{2-x} have shown [295] Figure 37, that E'_g decreases rapidly over the range $2 \leq x \leq 1.5$ and at a slower (linear) rate over the range $0 \leq x \leq 1.5$ so that $E'_g \simeq 1.1$ eV in $SnSe_2$. Similarly the birefringence Δ ($\lambda = 1$ μm) whose extrapolated value is ca. 0.65 in SnS_2 decreases to 0.38 in $SnSe_2$ [297].

The spectral variation of $\varepsilon_2(\omega)$ in $SnSe_2$(290, 77 K) has been determined [297] over the range 0.05 to 3.7 eV from separate measurements of α and $n_\perp$, Figure

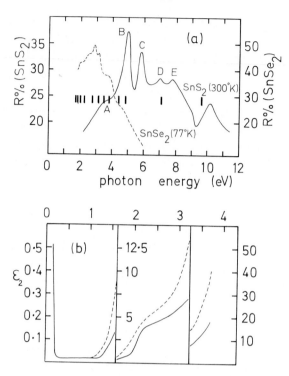

Fig. 37a–b. (a) The reflection spectra ($\mathbf{E}\perp\mathbf{c}$) of SnS$_2$ (300 K) [292] and SnSe$_2$ (77 K) [293]. Absolute values of $R\%$ for SnSe$_2$ are unreliable since another sample gave values of $R\%$ ca. 20% less than those shown. Vertical lines mark reported [293] positions of reflection peaks in SnSe$_2$ (77 K). (b) The ε_2 spectrum ($\mathbf{E}\perp\mathbf{c}$) of SnSe$_2$ [297] at 290 K (dashed line) and 77 K (full line).

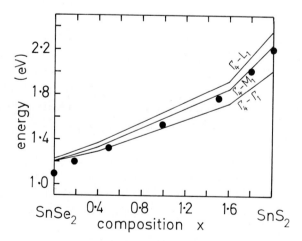

Fig. 37c. Calculated variation of lowest energy gaps with composition x in SnS$_x$Se$_{2-x}$ [298] – solid circles represent experimental values [295].

37. The absorption edge has been analysed [297] in terms of forbidden indirect transitions across energy gaps (ignoring phonon energy) $E_g'^{(1)} = 1.03$ eV, $E_g'^{(2)} = 1.30$ eV and allowed direct transitions across an energy gap $E_g = 1.97$ eV (77 K). The direct transitions give rise to the characteristic edge in $\varepsilon_2(\omega)$, Figure 37, at ca. 2 eV which appears as an oscillation in the wavelength modulated transmission spectrum [299]. Other SnSe$_2$ absorption edge measurements have been interpreted in terms of forbidden indirect transitions across an energy gap E_g' (300 K) = 1.0 eV [296, 300, 301]. 0.97 eV [294], 1.1 eV [295] and forbidden direct transitions across $E_g = 1.62$ eV [294].

The electron energy band structures of SnS$_2$, SnSe$_2$ have been calculated by 'a priori' [298, 302] and 'empirical' [303, 304] pseudo potential methods. The SnS$_2$ band structure shown in Figure 38 [302] identified the (composite) absorption threshold with direct $\Gamma_4 \to \Gamma_1$ transitions at 2.01 eV and indirect transitions $\Gamma_4 \to M_1$ at 2.24 eV and $\Gamma_4 \to L_1$ at 2.35 eV, all these transitions being forbidden [304] for $\mathbf{E} \perp \mathbf{c}$. The calculated variation in these lowest transition energies with composition x in SnS$_x$Se$_{2-x}$ [298], Figure 37, is similar to the observed variation in band gap energy [295] and supports the experimental conclusions that the absorption thresholds in SnS$_2$ and SnSe$_2$ are due to a number of neighbouring transitions. At higher energies structure in the calculated $(\varepsilon_2)_\perp$ spectrum [304] of SnS$_2$ matches that of the reflectivity spectrum, Figure 37. Most of the structure is a composite of neighbouring interband transitions, or arises from volume effects, with the possible exception of the edge A_1 at 3.8 eV which is identified with an M_0 $\Gamma_3' \to \Gamma_1$ transition [304], $\Gamma_5 \to \Gamma_1$ in the notation of Figure 38. On the calculated SnSe$_2$ band structure [304, 298] this transition produces an M_0 edge at ca. 2.0 eV which is close to the observed position, Figure 37.

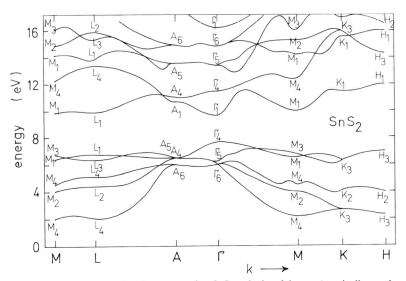

Fig. 38. An electron energy band structure for SnS$_2$ calculated by an 'a-priori' pseudo potential method [302].

10.7. GROUP V CHALCOGENIDES

The chalcogenides of the Group V B elements As, Sb, Bi will be considered.

As_2S_3, As_2Se_3

As_2S_3 (occurring naturally as the mineral orpiment) and As_2Se_3 crystals are isomorphic [305]. The As_2S_3 crystal structure consists of As_2S_3 layers stacked along the b axis of the monoclinic unit cell, the dimensions of the tetramolecular (two layer) unit cell are $a = 11.47$, $b = 9.57$, $c = 4.24$ Å; $\beta = 90°\ 27'$ [306]. For most purposes the unit cell can be taken as orthorhombic since the ac angle $\beta \simeq 90°$. As_2S_3 (and As_2Se_3) crystals show pronounced micaceous (010) cleavage across the Van der Waals bonds between adjacent layers, the alignment of adjacent layers being largely determined by packing considerations.

The orpiment structure contains centres of symmetry located on the mid planes between layers (and which transform one layer into another) with the result that the infra-red and Raman scattering frequencies are mutually exclusive. In the limiting case of zero layer–layer interaction however the absence of inversion symmetry in the single layer factor group would allow vibration modes to be both infra-red and Raman active. Also on the single layer model the thirteen fundamentals in the restrahlen spectra are separated to give seven restrahlen bands for $\mathbf{E}\|\mathbf{c}$ and six for $\mathbf{E}\|\mathbf{a}$. With weak layer–layer interaction the 30 doubly degenerate normal modes of the single layer split to give the 60 normal modes of the two layer (20 atom) crystal unit cell; the result is a set of closely spaced doublets. This interlayer interaction splitting (or Davydov splitting) is observed in As_2S_3, As_2Se_3 [307]; a *layer* vibration possessing both Raman and infra-red activity gives rise to two *crystalline* offspring, one having all the Raman strength and the other all the infra-red strength. The largest optical phonon Davydov splitting (i.e. the frequency difference between a line in the infra-red spectrum and its partner seen in the Raman spectrum) is 6 cm^{-1} for As_2S_3 and 5 cm^{-1} for As_2Se_3. Expressed as a fraction of the total (molecular) frequency the rms splitting is about 2% for both crystals, this gives some indication of the layer–layer interaction strength in these crystals.

Normal incidence transmission and reflection measurements on cleaved single crystals of As_2S_3 and As_2Se_3 give spectra for $\mathbf{E}\|\mathbf{c}$, $\mathbf{E}\|\mathbf{a}$. Figure 39 shows the dispersion of the principal refractive indices n_a ($\mathbf{E}\|\mathbf{a}$), n_c ($\mathbf{E}\|\mathbf{c}$) at the absorption threshold in As_2S_3 together with n_b as found from the measured birefringence ($n_a - n_b$) [308]. The absorption edge for $\mathbf{E}\|\mathbf{c}$ in As_2S_3 occurs to the low energy side of that for $\mathbf{E}\|\mathbf{a}$ [308, 309] both edges rising to structure labelled A, B, C, Figure 39, for $\mathbf{E}\|\mathbf{c}$ and A', B', C' ($\mathbf{E}\|\mathbf{a}$) at energies (77 K) A (2.89 eV), B (2.99 eV), C (3.17 eV); A' (2.95 eV), B' (3.05 eV), C' (3.235 eV) [308]. This structure (which sharpens and moves to higher energies with decreasing temperature) is also observed in the reflection spectra [308, 310]. It has been suggested [310] that A, B, C are each the first member of a two-dimensional exciton series associated

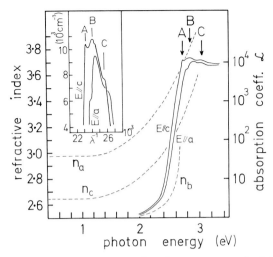

Fig. 39. Full lines show the absorption edges for $\mathbf{E}\|\mathbf{c}$, $\mathbf{E}\|\mathbf{a}$ in single crystals of As_2S_3 (290 K). Dashed lines show the dispersion of the refractive indices $n_a(\mathbf{E}\|\mathbf{a})$, $n_c(\mathbf{E}\|\mathbf{c})$ and $n_b(\mathbf{E}\|\mathbf{b})$. Inset shows in greater detail the appearance of absorption bands A, B, $C(\mathbf{E}\|\mathbf{c})$ and A', B', C' ($\mathbf{E}\|\mathbf{a}$) in As_2S_3 (77 K) [308]. (Reprinted by permission of the Royal Society)

with a valence sub-band and that the doublets AA', BB', CC' (having approximately the same energy separation of 0.06 eV) are due to interlayer interaction (Davydov) splitting.

The absorption thresholds in As_2S_3, which would seem to be built up from the tails of A and B ($\mathbf{E}\|\mathbf{c}$), can be described by equations of the form $\alpha = a(\hbar\omega + b)^2$ [308, 311, 312, 313] but the similarity of this expression to that for indirect transitions, (35), may be accidental.

In As_2Se_3 also the absorption threshold for $\mathbf{E}\|\mathbf{c}$ occurs to the low energy side of that for $\mathbf{E}\|\mathbf{a}$, both edges show structure at 2.01 and 2.19 eV (10 K) [314].

At energies above threshold the absorption spectra of As_2S_3, As_2Se_3 (room temperature) are as shown in Figure 40 [315] derived by Kramers-Kronig analysis of the reflectivity spectra. For both compounds the absorption rises first for $\mathbf{E}\|\mathbf{c}$, Figure 39, but is then overtaken by α for $\mathbf{E}\|\mathbf{a}$. All the spectra of Figure 40 show the first edge at 2–3 eV and a second prominent edge at 7–8 eV. The number of valence electrons n contributing to the absorption has been calculated from measured $\varepsilon_2(\omega)$ by a sum rule, the graph of n versus $\hbar\omega$ showing a plateau (where $n \simeq 3$ electrons atom^{-1}) around 8 eV for both compounds, c.f. Figure 42(c). Since the number of s and p electrons per atom is 5.6 i.e. $(2\times 5 + 3\times 6)/(2+3)$ of which 2.4 are bonding and 3.2 non-bonding so the two principal edges of Figure 40 are attributed to the onset of transitions involving non-bonding v.b states respectively [315].

The spectral variation of ε_2 of As_2S_3 (room temperature) for $\mathbf{E}\|\mathbf{a}, \mathbf{b}, \mathbf{c}$ has been derived from electron energy loss measurements in the range 3 to 35 eV [316], Figure 41. Anisotropy affects the amplitude rather than the energy of the

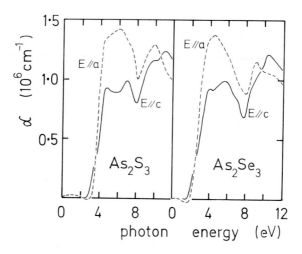

Fig. 40. Absorption spectra $E\|c$ (full line), $E\|a$ (dashed line) for As_2S_3 and As_2Se_3 (room temperature) [315]. The negative values at low photon energies are a measure of the experimental errors.

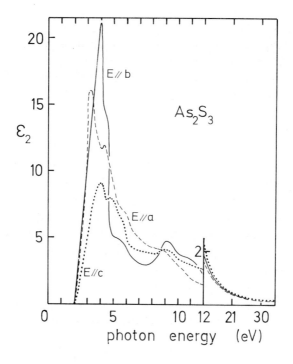

Fig. 41. ε_2, the imaginary part of the dielectric constant of As_2S_3 for $\mathbf{E}\|\mathbf{a}$ (dashed line), $\mathbf{E}\|\mathbf{b}$ (full line) and $\mathbf{E}\|\mathbf{c}$ (dotted line [316].

structure occurring in the range 2 to 12 eV, no further structure occurs in the range 12 to 30 eV and the derived graph of $n(\omega)$ rises smoothly to a value of 6 electrons per atom at this value.

Sb_2S_3, Sb_2Se_3

Sb_2S_3, which occurs naturally as the mineral stibnite, and Sb_2Se_3 are layer-type orthorhombic crystals, D_{2h}^{16} with tetramolecular unit cell dimensions (Å) $a = 11.229, 11.58$; $b = 11.310, 11.68$; $c = 3.839, 3.98$ respectively [119]. The complex structure of these compounds, in which the atoms are arranged to form thread-like molecules, has been explained by a mesomeric bonding system [290] which couples neighbouring thread molecules to form bands which are combined into layers parallel to (010) planes. Because of the comparatively few weak mesomeric bonds between layers both compounds show pronounced cleavage yielding mirror-like (010) surfaces which show signs of oxidation after exposure to the atmosphere for some days.

The infra-red reflectivity of Sb_2S_3 (300 and 95 K) over the range 70 to 500 cm^{-1} has been analysed to give the optic mode frequencies appropriate to $\mathbf{E}\|\mathbf{c}$, $\mathbf{E}\|\mathbf{a}$ [317]. At higher photon energies, but still below the absorption threshold, nearly all the dispersion measurements on Sb_2S_3 have been made on evaporated films although single crystals are reported to be biaxial negative [318]. One set of evaporated film measurements [319] finds n, $\varkappa$ to be independent of film thickness (contrary to other measurements [320]) with n increasing from ca. 2.15 ($\lambda = 24\ \mu m$) to 2.57 ($\lambda = 5\ \mu m$). The refractive index seems to depend upon evaporation conditions since other investigators give $n(\lambda = 5\ \mu m) = 2.7$ [321], 2.4 [322], 2.62 [318] compared with a single crystal value of 2.8 [323]. The onset of intrinsic absorption in Sb_2S_3 (300 K) occurs in the vicinity of 1.7 eV, the crystal absorption threshold ($\alpha < 50$ cm^{-1}) showing an extended long wavelength tail. For $\alpha > 50$ cm^{-1} the absorption coefficient $\alpha_\perp$ ($\mathbf{E}\perp\mathbf{c}$) varies exponentially with $\hbar\omega$ [324] but does not apparently obey Urbachs rule since the slope of the $\ln \alpha_\perp$ versus $\hbar\omega$ graph is almost constant over the temperature range 0 to 70°C. Over this temperature range the absorption edge (at $\ln \alpha_\perp = 5$) ranges from $\hbar\omega \approx 1.77$ eV to $\hbar\omega = 1.64$ eV. Taking the energy at which $\ln \alpha = 5$ as a measure of E_g the (linear) graphs of E_g versus $T°C$ for $\mathbf{E}\perp\mathbf{c}$, $\mathbf{E}\|\mathbf{c}$ show a change of slope around $T = 18°C$ during which dE_g/dT changes from -7×10^{-4} eV °C^{-1} ($T < 16°C$) to -6×10^{-4} eV °C^{-1} [324]; this behaviour is taken as evidence for a second order phase change in Sb_2S_3 [324, 325].

An electron energy band structure for Sb_2S_3 calculated by the LCAO method [326], neglecting layer–layer interaction, gives a direct energy gap of 2.2 eV located at $k = 0.85\ \pi/c$. The selection rules for optical dipole transitions are (a) $\mathbf{E}\|\mathbf{c}$; $\Gamma_1 \to \Gamma_2$, $\Gamma_5 \to \Gamma_6$, $\Delta_1 \to \Delta_1$, $\Delta_2 \to \Delta_2$. (b) $\mathbf{E}\perp\mathbf{c}$; $\Gamma_1 \to \Gamma_6$, $\Gamma_2 \to \Gamma_5$, $\Delta_1 \to \Delta_2$.

In the case of Sb_2Se_3 transmission and reflection measurements, $\mathbf{E}\perp\mathbf{c}$ [327] identify the absorption edge with direct transitions across an energy gap E_g (300 K) ≈ 1 eV; the edge rises to an absorption peak at 1.18 eV [328]. Other

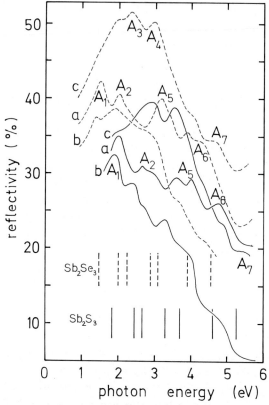

Fig. 42a. The reflectivity spectra of Sb_2S_3 (full line) and Sb_2Se_3 (dashed line) at 90 K for $\mathbf{E}\|\mathbf{a}$, $\mathbf{E}\|\mathbf{b}$, $\mathbf{E}\|\mathbf{c}$ – denoted by a, b, c respectively [332]. For clarity the curves have been shifted downwards relative to the reflectivity scale: for Sb_2S_3 – curve c down by 9%, curve a by 10% curve b by 7%; for Sb_2Se_3 – curve c down by 6%, curve a by 8%, curve b by 9%. The vertical lines mark the positions of structure A_1–A_7 in Sb_2S_3 (full lines) and Sb_2Se_3 (dashed lines).

measurements identify the absorption threshold in Sb_2Se_3 with indirect transitions across an energy gap E'_g of 1.16 eV [329] or 0.89 eV [330].

Figure 42(a) shows the reflectivity spectra of Sb_2S_3 and Sb_2Se_3 (90 K) over the range 0.7 to 5.5 eV for the polarizations $\mathbf{E}\|\mathbf{a}, \mathbf{c}, \mathbf{b}$, the latter spectra being obtained from the polished and etched surface cut perpendicular to the cleavage plane [332]. The reflection spectra of Sb_2S_3 and Sb_2Se_3 are similar. For each compound the $\mathbf{E}\|\mathbf{a}$, $\mathbf{E}\|\mathbf{b}$ spectra are identical apart from differences attributable to the polished surface. Comparing the $\mathbf{E}\|\mathbf{a}$ or $\mathbf{b}$ spectrum with that for $\mathbf{E}\|\mathbf{c}$ it is seen that peak A_1 is absent in the $\mathbf{E}\|\mathbf{c}$ spectrum and peaks A_3, A_4 absent from the $\mathbf{E}\|\mathbf{a}, \mathbf{b}$ spectrum. The polarization dependence of peak A_1 suggests that this peak is associated with transitions along one of the two-fold crystal axes [332], e.g. the Λ axis. There have not been any joint density of states calculations for Sb_2S_3, Sb_2Se_3 and the remaining reflection peaks A_2, A_3 etc. are unidentified.

Over the range 5 to 25 eV the normal incidence reflectivity of Sb_2Se_3 (room temperature) has been measured for $\mathbf{E}\|\mathbf{c}, \mathbf{E}\|\mathbf{a}$ [331], Figure 42(b), while the

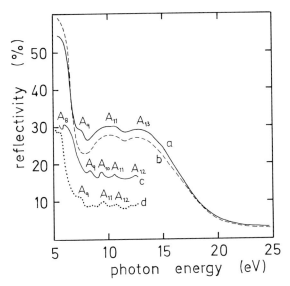

Fig. 42b. The room temperature reflectivity of Sb_2S_3 over the range 5 to 13 eV, full line curve c, for unpolarized light $\mathbf{E}\perp\mathbf{c}$ [332]. The room temperature reflectivity of Sb_2Se_3 over the range 5 to 25 eV for $\mathbf{E}\|\mathbf{c}$, full line curve a, and $\mathbf{E}\|\mathbf{a}$, dashed line curve b. [321]. The room temperature reflectivity of Sb_2Se_3 over the range 5 to 13 eV, dotted curve d, for unpolarized light, $\mathbf{E}\perp\mathbf{c}$ [332], for clarity this curve has been shifted downwards by 10%.

reflectivity of Sb_2Se_3 and Sb_2S_3 (295 K) has been measured for unpolarized light ($\mathbf{E}\perp\mathbf{c}$) over the range 5 to 13 eV [332], Figure 42(b). A Kramers-Kronig analysis of the Sb_2Se_3 reflectivity data [331] gives the $\varepsilon_2(\omega)$ spectra shown in Figure 42(c) and hence $n(\omega)$ for both polarizations; $n(\omega)$ is the number of electrons per molecule participating in transitions up to frequency ω. Both the $n(\omega)$ curves of Figure 42(c) show an inflection in the vicinity of 8 eV which suggests, c.f. As_2S_3, As_2Se_3, that the preceding transitions mainly involve the non-bonding p electrons [333, 334]; this low energy portion of the $n(\omega)$ curve looks as if it might saturate around 16 electrons per molecule (3.2 electrons atom^{-1}). Above 8 eV the $n(\omega)$ curves rise to ca. 25 electrons per molecule near 24 eV and will presumably saturate at 28 (valence) electrons per molecule at $\hbar\omega > 24$ eV; this higher energy region ($8 < \hbar\omega \leq 24$ eV) is evidently associated with transitions involving the bonding valence electrons.

Sb_2Te_3 is rhombohedral, D_{3d}^5, or hexagonal on a larger unit cell ($c =$, $a = 4.25$ Å [119]), and is isostructural with Bi_2Se_3, Bi_2Te_3. All three M_2X_3 compounds consist of five-layer sheets arranged perpendicular to the three-fold crystal axis where, in each sheet, the atomic layers are arranged in the sequence $X^{(1)}-M-X^{(2)}-M-X^{(1)}$; the superscripts identify the environment of the Te(Se) atoms. Weak bonding between the $X^{(1)}$ atoms in adjacent sheets explains the easy cleavage perpendicular to the crystal c axis.

Reflection measurements ($\mathbf{E}\perp\mathbf{c}$) on cleaved p-type single crystal Sb_2Te_3 [327,

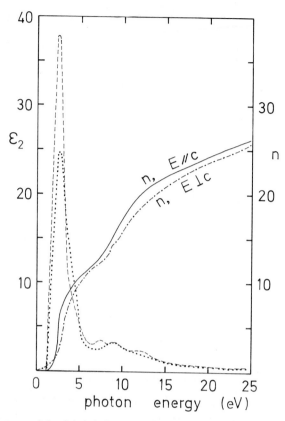

Fig. 42c. Imaginary part of the dielectric function of Sb_2Se_3 (room temperature) for $\mathbf{E} \| \mathbf{c}$ (dashed line) and $\mathbf{E} \| \mathbf{a}$ (dotted line). Also shown is the associated variation of $n(\omega)$ – the number of electrons per molecule participating in transitions up to frequency ω – as calculated from a sum rule [331].

346] have shown that the reflectivity R increases rapidly with λ for $\lambda > 10\ \mu m$ due to free carrier absorption. With decreasing λ below the reflection minimum at $\lambda \approx 9\ \mu m$ R again increases due to interband absorption. At the interband absorption threshold the spectral variation of α (allowing for free carrier absorption) has been described [346] in terms of indirect transitions across an energy gap E'_g (300 K) = 0.21 eV. Measurements on the alloy system [346] $Sb_{2X}Bi_{2(1-X)}Te_3$ show that E'_g decreases non-linearly with X in the range $0.8 < X < 1$ and linearly over the remaining composition range to $E'_g = 0.13$ eV at $X = 0$.

The 90 K reflection spectrum ($\mathbf{E} \perp \mathbf{c}$) of a cleaved Sb_2Te_3 crystal is shown in Figure 43 for the range 0.7 to 5.5 eV together with the 290 K reflection spectrum over the range 5.5 to 12.5 eV. The Sb_2Te_3 spectrum is similar to that of the isostructural compounds Bi_2Se_3, Bi_2Te_3, also shown in Figure 43, and the spectra of all three compounds will be described later (see Bi_2Te_3). The reflection peak labelling scheme of rhombohedral Sb_2Te_3, Figure 43, follows that used for orthorhombic Sb_2S_3, Sb_2Se_3, Figure 42, and plotting reflection peak positions (eV) against lattice parameter (viz. c(Å) for Sb_2S_3, Sb_2Se_3 and 'a' (Å) for Sb_2Te_3)

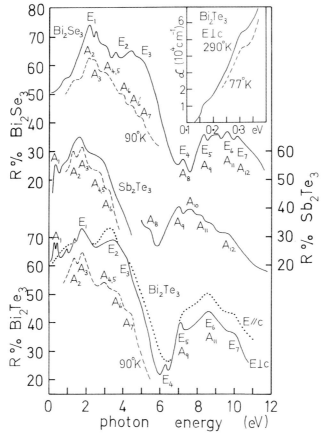

Fig. 43. The room temperature reflectivity spectra, $\mathbf{E} \perp \mathbf{c}$, of single crystal Bi_2Se_3 [341], Sb_2Te_3 [342] and Bi_2Te_3 [341] over the range 0.1 to 12 eV. Also shown (dotted) is the reflection spectrum, $\mathbf{E} \| \mathbf{c}$, of a Bi_2Te_3 crystal surface cut, and polished, perpendicular to the cleavage planes – absolute values of $R\%$ for this surface may be unreliable [341]. The dashed curves show the reflectivity spectra, $\mathbf{E} \perp \mathbf{c}$, of single crystal Bi_2Se_3, Sb_2Te_3 and Bi_2Te_3 [342] (90 K) and the inset shows the fundamental absorption edge, $\mathbf{E} \perp \mathbf{c}$, in Bi_2Te_3 [341].

demonstrates the correspondence of the spectra; each reflection peak moves to higher energy in the sequence Sb_2Te_3, Sb_2Se_3, Sb_2S_3 and the change in symmetry (between Sb_2Te_3 and Sb_2Se_3) leads to the splitting of peak $A_{4,5}$ and the polarization dependence of the reflection spectra.

Despite the dissimilar structures of Sb_2Se_3, Sb_2Te_3 solid solutions having composition in the range between Sb_2Te_3 and Sb_2TeSe_2 can occur which have a single phase hexagonal structure; the lattice parameter 'a' decreases from ca. 4.27 to 4.12 Å and 'c' decreases from ca. 30.4 to 29.5 Å with increasing Sb_2Se_3 content [347]. Transmission measurements on these single crystal alloys indicate that E_g decreases with increasing Sb_2Se_3 content [327]; the room temperature alloy reflection spectrum did not show peaks A_2, A_3 etc.

Bi_2S_3 has an orthorhombic crystal structure, tetramolecular unit cell dimensions

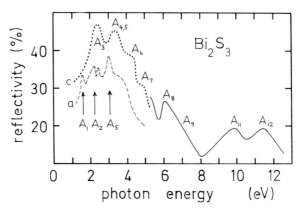

Fig. 44. The reflectivity spectra of Bi_2S_3 (77 K) for $\mathbf{E}\|\mathbf{c}$, and $\mathbf{E}\|\mathbf{a}$ over the range 1 to 5 eV [340]. Also shown is the unpolarized light reflectivity spectrum (full line) of Bi_2S_3 (293 K) over the range 5 to 12.5 eV [340].

(Å), [119], $a = 11.13$, $b = 11.27$, $c = 3.97$, space group D_{2h}^{16}, and is isomorphic with Sb_2S_3, Sb_2Se_3. The cleaved single crystal (77 K) normal incidence reflection spectra for $\mathbf{E}\|\mathbf{c}$, $\mathbf{E}\|\mathbf{a}$ over the range 1 to 5 eV is shown in Figure 44 together with the unpolarized light reflection spectrum of Bi_2S_3 (293 K) over the range 5 to 12.5 eV [340]. The reflection peak labelling scheme follows that used for Sb_2S_3, Sb_2Se_3 whose reflection spectra are shown in Figure 42. At 77 K the Bi_2S_3 spectrum ($\mathbf{E}\|\mathbf{c}$) shows two strong maxima at 2.4 (A_3) and 3.1 eV ($A_{4,5}$) with weaker structure at ca. 1.8, 4.0 (A_6) and 4.7 eV (A_7). For $\mathbf{E}\|\mathbf{a}$ a strong maximum occurs at 3.05 eV (A_5) with weaker peaks at 1.65 (A_1), 1.9, 2.1 (A_2), 2.3 and 4.0 eV. The unpolarized light spectrum (293 K) shows peaks at ca. 6 (A_8), 7 (A_9), 9.9 (A_{11}) and 11.3 eV (A_{12}). With this labelling scheme the polarization of the reflection peaks in Bi_2S_3 is the same as in Sb_2S_3, Sb_2Se_3; the energy position of each A peak shows an approximately linear shift to lower energy with increasing unit cell dimension 'c' i.e. in the sequence Sb_2S_3, Sb_2Se_3, Bi_2S_3, [340].

Bi_2Se_3, Bi_2Te_3 are isostructural with Sb_2Te_3, point group D_{3d}, and have hexagonal unit cell dimensions $a = 4.14$, $c = 28.6$ Å [348] and $a = 4.35$, $c = 30.3$ Å respectively.

The normal incidence infra-red reflectivity ($\mathbf{E}\perp\mathbf{c}$) from the cleaved surfaces of Bi_2Se_3, Bi_2Te_3 (room temperature) has been measured over the range 10 to 280 cm^{-1} on samples of low free carrier density [335, 336]. Both lattice vibration spectra (normally swamped by the free carrier reflectivity) can be described by a two-oscillator model having eigenfrequencies of 92 and 69.5 cm^{-1} for Bi_2Se_3 and 100.8 and 49.3 cm^{-1} for Bi_2Te_3. The lattice force constants of Bi_2Se_3 and Bi_2Te_3 are approximately the same but the static dielectric constants $\varepsilon_\perp(0)$ of 100 ± 10 and 360 ± 50 respectively demonstrate the larger polarizability of the Te atom.

Like Sb_2Te_3 so Bi_2Se_3, Bi_2Te_3 are degenerate semiconductors showing strong free carrier absorption which has to be allowed for, together with the Burstein shift, in determining the interband absorption edge.

Transmission and reflection measurements ($\mathbf{E} \perp \mathbf{c}$) on n-type single crystals of Bi_2Se_3 whose carrier concentration was changed from 10^{19} cm^{-3} to 10^{18} cm^{-3} by annealing in Se vapour give a corrected optical band gap $E_0 = 0.21 \pm 0.01$ eV [349] where $|-dE_0/dT| < 10^{-4}$ eV K^{-1}.

Single crystals of Bi_2Te_3 are normally p-type but can be made intrinsic or n-type by doping with iodine; transmission measurements on a nearly intrinsic crystal [350] showed that the fundamental absorption edge ($\alpha < 2500$ cm^{-1}) resembled that due to indirect transitions across an energy gap $E'_g = 0.13$ eV (290 K) where $dE'_g/dT = -0.95 \times 10^{-4}$ eV K^{-1}. The inset to Figure 43 shows the absorption edge ($\mathbf{E} \perp \mathbf{c}$) in single crystal p-type Bi_2Te_3 at room temperature [341]; there is an inflection at 0.18 eV which may be due to the lowest direct interband transition and a further absorption edge occurs at 0.3 eV, both these features are observed in the Bi_2Te_3 reflection spectrum.

Transmission measurements ($\mathbf{E} \perp \mathbf{c}$) on single crystal alloys of Bi_2Te_3 and Bi_2Se_3 [341, 351], which are isostructural with Bi_2Te_3, have shown that, with increasing Bi_2Se_3 content, E'_g (300 K) increases from ca. 0.145 eV in Bi_2Te_3 to 0.295 eV in Bi_2Te_2Se and then, allowing for free carrier effects [341], decreases to 0.167 eV in Bi_2Se_3. The single crystal reflection spectra ($\mathbf{E} \perp \mathbf{c}$) of Bi_2Te_3, Bi_2Se_3 (300 K) are shown in Figure 43; the principal peaks are labelled E_1—E_7. Reflection measurements on the single crystal alloys [341] have shown that in going from Bi_2Te_3 to Bi_2Se_3 peaks E_1 to E_7 show an identical linear shift to higher energies over the composition range Bi_2Te_3 to Bi_2Te_2Se. Above this composition the shifts are again equal but very small, the peaks occurring in almost the Bi_2Se_3 positions. This systematic shift in energy of the reflection peaks E_1 to E_7, and also of E'_g, illustrates the similarity between the band structures of Bi_2Se_3, Bi_2Te_3 and can also be taken as evidence [341] for the proposed bonding model of Bi_2Te_3 [352]. In this model the Bi—Te$^{(1)}$ bond in the five-layer sheet Te$^{(1)}$ BiTe$^{(2)}$ BiTe$^{(1)}$, c.f. Sb_2Te_3, is considered to be more ionic, and stronger, then the Bi—Te$^{(2)}$ bond; substituted Se atoms will first replace the, less electronegative, weakly bound Te$^{(2)}$ atoms thereby increasing the ionicity of the average BiTe$^{(2)}_x$ Se$^{(2)}_{1-x}$ bond and so increasing the energy gap. Hence the increase in E'_g and shift in peaks E_1—E_7. When all Te$^{(2)}$ sites are occupied, as in Bi_2Te_2Se, further Se atoms will go into Te$^{(1)}$ sites, at random, and by attracting charge into Se$^{(1)}$—Bi make the Bi—Se$^{(2)}$ bond less ionic; this could account for the observed decrease in energy gaps over the composition range Bi_2Te_2Se to Bi_2Se_3.

The reflectivity spectra of the three isostructural compounds Sb_2Te_3, Bi_2Se_3, Bi_2Te_3 given in Figure 43 are very similar to one another. At low temperatures, where the spectra show more structure, corresponding peaks in the three spectra can be similarly labelled (A_1, A_2 etc). If the energies of each identically labelled peak, e.g. A_3, are plotted against the rhombohedral lattice parameter (which increases in the sequence Bi_2Se_3, Sb_2Te_3, Bi_2Te_3) then, as evident from Figure 43, the peak in Sb_2Te_3 occurs to the low energy side of that in Bi_2Te_3 whereas the lattice parameter and Bi_2Se_3—Bi_2Te_3 alloy measurements given earlier suggest that the Sb_2Te_3 peak positions should be intermediate between that of Bi_2Se_3 and Bi_2Te_3. This behaviour is attributed [342] to the different relativistic corrections in Bi and Sb, with decreasing atomic number the energy of the s-states increases more than that of the p-states; this may identify the more markedly displaced peaks A_8 to A_{12} (E_5 to E_7) with s-like valence states.

As in the case of As_2S_3, As_2Se_3 the 28 valence electrons per unimolecular cell are distributed in two groups of valence bands, well separated in energy [341, 342]. In the case of Bi_2Te_3 [338] and, presumably, Bi_2Se_3, Sb_2Te_3 the spectral variation of $n(\omega)$, c.f. Figure 42(c), first saturates at $n = 14$ electrons per molecule around $\hbar\omega = 6$ eV, corresponding to the refléctivity minimum in the region of E_4, Figure 43. Transitions from the lower s-like (see previously) v.b states to the p-like c.b then produce the second increase in reflectivity containing bands A_9, A_{10} etc. (E_5, E_6, E_7).

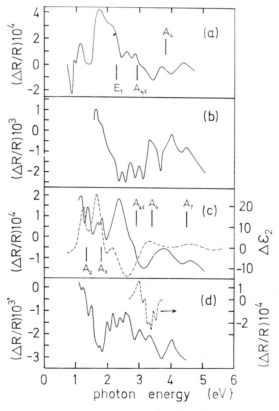

Fig. 45. (a) The electroreflectance spectrum of Bi_2Se_3 (room temperature) for $\mathbf{E} \perp \mathbf{c}$ [337]. (b) The thermoreflectance spectrum of Bi_2Se_3 (120 K) for $\mathbf{E} \perp \mathbf{c}$ [337]. (c) The electroreflectance spectrum (full line) of Bi_2Te_3 for $\mathbf{E} \perp \mathbf{c}$ [337]. The dashed curve represents the field induced change in the imaginary part of the dielectric constants of Bi_2Te_3 (room temperature) [338]. (d) The thermoreflectance spectrum (full line) of Bi_2Te_3 (120 K) for $\mathbf{E} \perp \mathbf{c}$ [337]. The dashed curve represents the thermoreflectivity of Bi_2Te_3 (80 K) [339]. The vertical lines labelled E or A mark the positions of structure in the reflectivity spectra, Figure 43.

Modulated reflectivity measurements over the range of the first broad reflectivity band (absorption threshold to $\hbar\omega = 6$ eV) are shown in Figure 45. In some cases the modulation spectrum reveals more structure than the direct spectrum. The electroreflectance spectrum of Bi_2Te_3 has identified peaks A_2, A_3, and $A_{4,5}$ of the reflectivity spectrum with two M_1 longitudinal saddle point edges and an M_0 threshold at the BZ centre [338]; the A_3 line is polarization dependent as is evident from the $\mathbf{E}\|\mathbf{c}$, $\mathbf{E}\perp\mathbf{c}$ spectra of Figure 43.

The electron energy band structure of Bi_2Te_3 has been calculated by pseudo potential methods, including spin-orbit splitting [343, 344], as shown in Figure 46. The band structure is very complicated and in the absence of calculated and

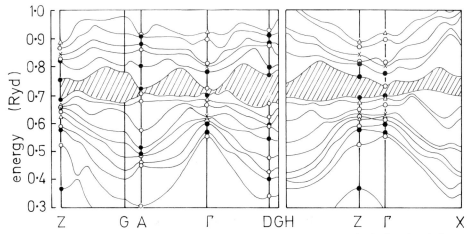

Fig. 46a. The band structure of Bi_2Te_3 along various symmetry lines. Hatched region indicates forbidden energy gap. Following symbols are used for notation of levels [343].

$\times = \Gamma_4^+ + \Gamma_5^+$ or $Z_4^+ + Z_5^+$,

$\triangle = \Gamma_4^- + \Gamma_5^-$ or $Z_4^- + Z_5^-$,

$\bullet = \Gamma_6^+, Z_6^+, A_5 + A_6$ or $D_5 + D_6$,

$\circ = \Gamma_6^-, Z_6^-, A_7 + A_8$ or $D_7 + D_8$.

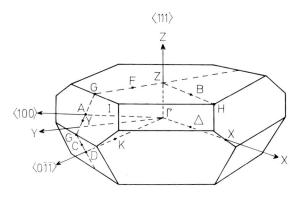

Fig. 46b. The first Brillouin zone at Bi_2Te_3 showing symmetry points, lines and planes.

TABLE VI

Energies (eV) of reflectivity peaks of Bi_2Te_3 at 293 K [342] and theoretical energies (eV) of interband transitions and their nature

A_2	A_3	$A_{4,5}$	A_6	A_7	A_8	A_9	A_{10}	A_{11}	A_{12}	Ref.
1.36	1.80	2.95	3.4	4.5	(6.1)	7.4	(7.9)	8.6	10.6	[342]
—	$Z_6^+ \to (Z_4^- + Z_5^-)$	$Z_6^+ \to Z_6^-$	$(Z_4^+ + Z_5^+) \to Z_6^-$	$\Gamma_6^- \to (\Gamma_4^+ + \Gamma_5^+)$	$Z_6^- \to Z_6^+$	$(Z_4^+ + Z_5^+) \to (Z_4^- + Z_5^-)$	$\Gamma_6^+ \to (\Gamma_4^- + \Gamma_5^-)$	$(Z_4^- + Z_5^-) \to (Z_4^+ + Z_5^+)$	$(Z_4^- + Z_5^-) \to Z_6^+$	[344]
1.8		2.5	3.6	4.7	6.1; 6.6	6.8	8.0	9.2	10.4	
		$\Gamma_6^+ \to \Gamma_6^-$	$(Z_4^+ + Z_5^+) \to Z_6^+$	$\Gamma_6^+ \to (\Gamma_4^+ + \Gamma_5^+)$					$Z_6^- \to Z_6^+$	
		2.9	3.9	4.1					9.5	
		$(Z_4^+ + Z_5^+) \to$								
		$(Z_4^- + Z_5^-)$								
		2.9								
$Z_6^+ \to Z_6^-$	$Z_6^- \to Z_6^+$	$Z_6^+ \to (Z_4^- + Z_5^-)$	$Z_6^+ \to Z_6^-$	$(\Gamma_4^- + \Gamma_5^-) \to \Gamma_6^+$	$Z_6^+ \to Z_6^-$	$Z_6^+ \to Z_6^-$		$(A_7 A_8) \to (A_5 + A_6)$		[343]
1.5	1.6; 1.9	2.7	3.3	4.1	6.1	7.0		8.6		
	$Z_6^+ \to Z_6^-$	$\Gamma_6^+ \to \Gamma_6^-$	$(D_5 + D_6) \to$	$(Z_4^+ + Z_5^+) \to Z_6^-$	$Z_6^- \to Z_6^+$	$(D_5 + D_6) \to$				
			$(D_7 + D_8)$			$(D_7 + D_8)$				
	1.7	2.9; 2.8	3.3	4.1; 4.4	6.2	7.0				
	$(Z_4^- + Z_5^-) \to Z_6^+$	$\Gamma_6^- \to \Gamma_6^+$	$\Gamma_6^+ \to \Gamma_6^-$	$Z_6^- \to (Z_4^+ + Z_5^+)$						
	1.7; 1.9	3.0	3.4	4.6						
	$\Gamma_6^+ \to \Gamma_6^-$	$Z_6^- \to (Z_4^+ + Z_5^+)$	$(A_7 + A_8) \to$	$(Z_4^+ + Z_5^+) \to Z_6^-$						
			$(A_5 + A_6)$							
	2.0; 2.2	3.1	3.5	4.7						

measured $\varepsilon_2(\omega)$ spectra the transitions giving rise to the structure in the reflectivity spectra cannot be identified with certainty. However the possible interband transitions giving rise to peaks A_2 to A_{12} inclusive in Bi_2Te_3 are given in Table VI [345].

10.8. TRANSITION METAL DICHALCOGENIDES

The compounds MX_2 to be considered are those, Table VII, which crystallize with a layer structure built up from the sandwiches $X—M—X$ in which each metal atom (M) is in 6-fold coordination with the chalcogen (X) atoms lying in hexagonally packed planes, Figure 47. In order to form the six equivalent $M—X$ bonds it has been shown [335] that hybridization of the M s, p, d atomic orbitals must occur; the trigonal prism arrangement (stacking sequence ABA, Figure 47(b)) gives stronger bonds than the octahedral arrangement (stacking ABC, Figure 47(a)) but the latter is favoured by closer packing of the X atoms. In some of the tellurides, Table VII, distorted coordinations occur.

In a sense each sandwich layer, Figure 47, is a complete entity since it has the chemical composition MX_2 and the bonding requirements of the M atom are satisfied; for this reason preliminary band structure calculations for these compounds were based upon the single sandwich model. A real crystal however is built up of weakly bound sandwich layers stacked one upon another, the different possible stacking arrangements giving rise to a number of stacking polytypes. When the M atom has octahedral coordination within the layer, Figure 47(a), then a stacking sequence of ABC/ABC (where the stroke denotes layer separation) gives the $1s—CdI_2$ (hexagonal 1T) structure of the Group IV dichalcogenides. When the M atom has trigonal prismatic coordination, Figure 47(b),

TABLE VII

Coordination configuration in some of the layer structure transition metal dichalcogenides. Bracketed values refer to the ionic radius (Å) of the M^{4+} or X^{2-} ions [370]

M	X_2	S_2 (1.84)	Se_2 (1.98)	Te_2 (2.21)
IV				
Ti	$3d^2\ 4s^2$ (0.68)	octahedral	octahedral	octahedral
Zr	$4d^2\ 5s^2$ (0.79)	octahedral	octahedral	octahedral
Hf	$4d^2\ 6s^2$ (0.78)	octahedral	octahedral	octahedral*
V				
V	$3d^3\ 4s^2$ (0.63)	octahedral	octahedral	dist. octahed.
Nb	$4d^4\ 3s^1$ (0.74)	trig. prism	trig. prism	dist. octahed.
Ta	$5d^3\ 6s^2$ (0.68)	{trig. prism / dist. octahed.}	{trig. prism / dist. octahed.}	dist. octahed.
VI				
Cr	$3d^5\ 4s^1$ (0.63^{3+})	—	—	—
Mo	$4d^5\ 5s^1$ (0.70)	trig. prism	trig. prism	{trig. prism / dist. octahed.}
W	$5d^4\ 6s^2$ (0.70)	trig. prism	trig. prism	dist. octahed.

* Existence doubtful.

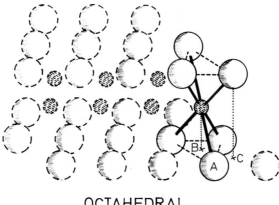

OCTAHEDRAL

Fig. 47a. An $X—M—X$ sandwich layer showing the octahedral coordination of the metal M atom (small circle) resulting from stacking ABC.

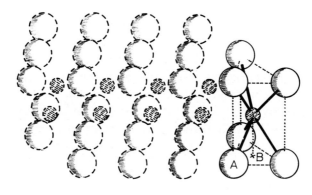

TRIGONAL PRISMATIC

Fig. 47b. An $X—M—X$ sandwich layer showing the trigonal prismatic coordination of the metal M atom (small circle) resulting from a stacking sequence ABA.

then the most commonly occurring stacking sequences are those shown in Figure 48; most of the Group V, VI dichalcogenides occur as one or more of these polytypes [354, 355, 356]. In some of the compounds of Table VII the metal atom is displaced to give a distorted coordination configuration, this results in buckled chalcogenide sheets and chains of metal atoms running through the structure [357, 358].

As in the case of the other layer compounds described earlier, e.g. As_2Se_3, some indication of the nature of the bonding in the crystal and the degree of interlayer attraction can be gained from infra-red and Raman measurements. These measurements have been restricted to the Group IV and VI layer compounds since the free carrier absorption in the Group V compounds prevents measurements in the infra-red.

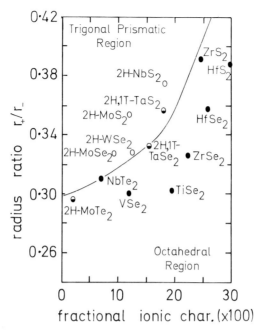

Fig. 47c. The radius ratio versus the fractional ionic character of the metal-chalcogen bond in various transition metal dichalcogenides [420].

○ Trigonal prismatic coordination.
● Octahedral coordination.
◐ Compounds exhibiting both configurations.

In the 2H—MoS$_2$ structure, Figure 48(d), the eighteen vibrational modes of the six atoms per unit cell have the following irreducible representation at the B.Z centre [359]

$$\Gamma = 2A_{2u} + B_{2g}^1 + B_{2g}^2 + B_{1u} + A_{1g} + 2E_{1u} + E_{2g}^1 + E_{2g}^2 + E_{2u} + E_{1g}.$$

The infra-red and Raman active mode vibrations are listed in Table VIII and described in Figure 49(a).

For the 1T—CdI$_2$ structure of the Group IV dichalcogenides there are three atoms per unit cell and the decomposition into irreducible representations at the Γ point is given by [361]

$$\Gamma = A_{1g} + E_g + 2A_{2u} + 2E_u.$$

These optic mode activities are listed in Table VIII and described in Figure 49(b).

The infra-red reflectivities of the following Group IV dichalcogenides have been measured [360]: – TiS$_2$; ZrS$_2$, ZrTe$_2$; HfS$_2$, HfSe$_2$, HfTe$_2$. Measurements were made on the cleaved single crystal surface (room temperature), $\mathbf{E} \perp \mathbf{c}$, so that only the E_u mode could be studied, Table 8. The spectra were of two types; ZrS$_2$, HfS$_2$ and HfSe$_2$ showed a strong restrahlen while TiS$_2$, ZrTe$_2$ and HfTe$_2$ showed

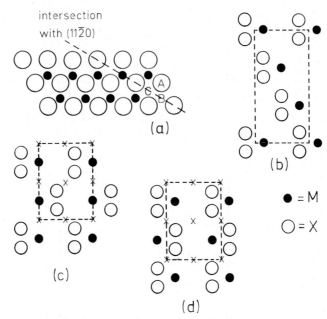

Fig. 48. (a) Plan view of the layer of Figure 47(b) in which the metal M atom has trigonal prismatic coordination. The dashed line shows the intersection with $(11\bar{2}0)$. (b) $(11\bar{2}0)$ section of the rhombohedral MoS$_2$ polytype, 3R—MoS$_2$. The stacking sequence is $ABA/BCB/CAC$. (c) $(11\bar{2}0)$ section of the 2H—NbS$_2$ polytype in which the stacking sequence is ABA/CBC. In this polytype the metal atoms lie directly above each other. (d) $(11\bar{2}0)$ section of the molybdenite, 2H—MoS$_2$, polytype; stacking sequence ABA-/BAB.

TABLE VIII

Symmetries and selection rules for the long wavelength acoustic and optic phonons in the 1T and 2H polytype geometries [360]

	Modes at $\mathbf{q}=0$ Irreducible representations	Acoustic	Optic	Selection rules
1T–polytypes D_{3d}	A_{1g}	0	1	Raman
	E_g	0	1	Raman
	A_{2u}	1	1	i.r(z) $\mathbf{E}\|\mathbf{c}$
	E_u	1	1	i.r(x, y) $\mathbf{E}\|\mathbf{c}$
2H–polytypes D_{6h}	A_{2u}	1	0	inactive
	B_{2g}^2	0	1	inactive
	A_{2u}	0	1	i.r(z) $\mathbf{E}\|\mathbf{c}$
	B_{2g}^1	0	1	inactive
	A_{1g}	0	1	Raman
	B_{1u}	0	1	inactive
	E_{1u}	1	0	inactive
	E_{2g}^2	0	1	Raman
	E_{1u}^1	0	1	i.r(x, y) $\mathbf{E}\perp\mathbf{c}$
	E_{2g}^1	0	1	Raman
	E_{1g}	0	1	Raman
	E_{2u}	0	1	inactive

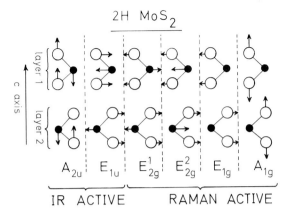

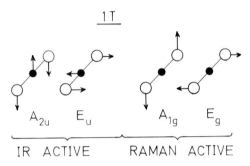

Fig. 49. Displacement vectors for the infra red and Raman-active modes in the 2H and 1T polytypes [360].

plasmon type reflections bands corresponding to free carrier densities in excess of 10^{20} cm^{-3}, Figure 50. Each restrahlen spectrum was described by a damped Lorentzian oscillator model which yielded the following TO (LO) phonon frequencies (cm^{-1}). ZrS$_2$—181 (350), HfS$_2$—166 (318), HfSe$_2$—98 (215). The oscillator strength of the restrahlen gave the following values for the effective charge parameter e_T^* [defined by $e_T^* = (\partial \bar{M}/\partial \bar{u})_E$, where $\bar{M}$ is the first order moment induced by the relative (or optical) displacements $\bar{u}$ of the atoms] viz. ZrS$_2$: $-e_T^* = 4.4e$, HfS$_2$: $-e_T^* = 3.9e$, HfSe$_2$: $-e_T^* = 4.3e$.

Raman scattering measurements on Group IV A dichalcogenide single crystals (300 K) [363] gave the Raman active mode frequencies listed in Table IX. The A_{1g} and E_{1g} mode energies are independent of the metal ion mass since this ion is stationary in an even Raman active mode, Figure 49. The mode frequencies scale almost as (chalcogen mass ratio)$^{-(1/2)}$ indicating that there is only a small variation in spring constant from compound to compound.

Of the Group VI dichalcogenides the infra-red reflection spectra have been measured for 2H—MoS$_2$ [362, 360], 2H—MoSe$_2$ [360, 364, 365], 2H—WS$_2$ [360], 2H—WSe$_2$ [360]. The restrahlen spectrum of all four compounds were

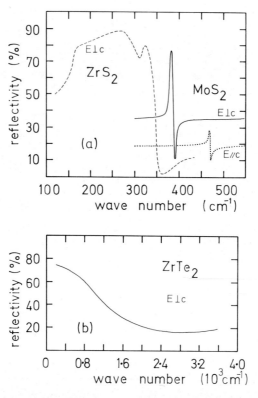

Fig. 50. (a) The reflectivity of 2H—MoS$_2$ single crystal (room temperature) for $\mathbf{E} \perp \mathbf{c}$ (full line) and $\mathbf{E} \| \mathbf{c}$ (dotted) [362]. The dashed line curve denotes the reflectivity of 1T—ZrS$_2$ single crystal (room temperature) for $\mathbf{E} \perp \mathbf{c}$ [360]. (b) The room temperature reflectivity spectrum of crystalline ZrTe$_2$* for $\mathbf{E} \perp \mathbf{c}$ [360].

similar, that for single crystal MoS$_2$ being shown in Figure 50. The $\mathbf{E} \| \mathbf{c}$, $\mathbf{E} \perp \mathbf{c}$ reflectivity spectra of 2H—MoS$_2$ shown in Figure 50 were obtained [362] by cutting the crystal to give a face parallel to the c axis; reflectivity measurements from the cleaved (00.1) surface only give the $\mathbf{E} \perp \mathbf{c}$ spectrum (as in MoSe$_2$ [364, 365). For MoSe$_2$, WS$_2$, WSe$_2$ the $\mathbf{E} \| \mathbf{c}$, $\mathbf{E} \perp \mathbf{c}$ restrahlen bands were obtained from

TABLE IX

Raman active mode frequencies (cm^{-1}) in TiSe$_2$, TiSe$_2$, ZrS$_2$, ZrSe$_2$, HfS$_2$ and HfSe$_2$ [363]

	S$_2$		Se$_2$	
	A_{1g}	E_g	A_{1g}	E_g
Ti	335	232	195	134
Zr	333	235	194	148
Hf	337	253	198	155

*Crystals were predominantly ZrTe$_3$, the CdI$_2$ phase does not apparently extend to the limit ZrTe$_2$ [366].

TABLE X

E_{1u} and A_{2u} phonon frequencies (cm^{-1})
in MoS$_2$, MoSe$_2$, MoTe$_2$, WS$_2$, WSe$_2$
(room temperature)

Compound	Phonon frequencies (cm^{-1})	
	$\nu(E_{1u})$	$\nu(A_{2u})$
MoS$_2$	384 [362]	470 [362]
MoSe$_2$	{288 [360], 277 [365]}	350 [360]
MoTe$_2$	240 [368]	
WS$_2$	356 [360]	435 [360]
WSe$_2$	245 [360]	305 [360]

the reflection spectrum of a pressed powder pellet [360], the weaker of the two bands being attributed to the small number of crystallites having the orientation **E**∥**c** in the pellet surface. The frequencies of the E_{1u} (**E**⊥**c**) and A_{2u} (**E**⊥**c**) optic modes, Table 8, obtained from a classical oscillator fit to the measured infra-red reflectivity are given in Table X.

The restrahlen bands of the Group VI compounds are very different to those of the Group IV compounds, Figure 50. Comparing MoS$_2$ and ZrS$_2$ for example (in which the atomic masses of Mo and Zr are nearly equal) the TO frequency is 384 cm^{-1} in MoS$_2$ and 181 cm^{-1} in ZrS$_2$ which implies a much larger force constant in the Group VI compound. The oscillator strength (i.e. effective charge e_T^* in MoS$_2$ is much smaller than in ZrS$_2$, Figure 50, which leads to the conclusion that the bonds in MoS$_2$ (and the other Group VI compounds of Table X) are largely covalent [362] whereas the Group IV semiconductors ZrS$_2$, HfS$_2$, HfSe$_2$, are mainly ionic [360] c.f. Figure 47(c).

Raman scattering measurements on the Group VI dichalcogenides have been restricted to 2H—MoS$_2$ [362, 367, 368, 369], MoSe$_2$ [364, 368], MoTe$_2$ [368]. In MoS$_2$ back scattered Raman lines were observed [362] at 287, 383 and 409 cm^{-1} at room temperature and assigned to the E_{1g}, E_{2g}^1 and A_{1g} modes respectively, Table 8. The E_{1u} infra-red active mode and the E_{2g}^1 Raman active mode differ only by an interlayer phase shift of π, Figure 49, and are degenerate in energy, as observed, for weak layer–layer interaction [359] as evidenced by easy cleavage between layers. The E_{2g}^2 mode, in which the molecular displacements in the unit cell correspond to rigid layer shear, occurs at 32 cm^{-1} [369, 367] in MoS$_2$. This frequency is about one tenth that of the dipolar mode (384 cm^{-1}) in the basal plane so that the interlayer bonding forces are about 100 times weaker than the intra-layer forces. Some of the features present in the second order (multiphonon) Raman spectrum have been identified as overtone or difference processes of the E_{1g}, E_{2g}^1 and A_{1g} modes [367].

The Raman spectra of MoSe$_2$, MoTe$_2$ [368] resemble that of MoS$_2$ in that weak interlayer forces lead to the near equality of the E_{1u} and E_{2g}^1 mode frequencies, Table X. The observed frequencies, MoTe$_2$ values bracketed, are $E_{1g} = 217$ (207), $A_{1g} = 361$ (321), $E_{2g}^1 = 285$ (237).

The infra-red measurements indicate that the Group IV semiconducting dichalcogenides are ionic, this is counter to the suggestion that the octahedral coordination, Table VII, results from the M—X hybrid bond configuration. For *ionic* compounds of the type MX_2 the preferred crystal structure is determined by the Madelung constant and the ratio r_+/r_- of the cation, anion radii. A structure, such as CdI$_2$, giving the required ligancy of 6 occurs for $0.65 > r_+/r_- > 0.33$ [371]; ratios greater or less than these limiting values give ligancies of 8 and 4 respectively. It can be argued however [372] that when the radius ratio lies close to one of the limiting values the compound may prefer to change the nature of its bonding rather than its ligancy and that this is why, Table VII, the Ti and Group VI compounds are non-ionic. The electronic configuration of Mo, W, Table VII, is such that they are able to form the $d^4 s p$ hybrid orbitals which overlap with a chalcogen p orbital to give the trigonal prismatic structure whereas the electron configuration of Ti ($3d^2 4s^2$) is different from that required ($d^2 s p^3$) for octahedral covalent bonding – the Ti compounds are apparently metallic.

The foregoing argument can be developed [420] by considering the fractional ionic character f_i of the compound defined by $f_i = 1 - \exp[-\frac{1}{4}(X_M - X_x)^2]$, where X_M, X_x are the electronegativities of the metal and chalcogen atom respectively. Plotting r_+/r_- against f_i, Figure 47(c), it is found that above the curved line compounds exhibit trigonal prismatic coordination while below the line only octahedral coordination is possible. Crystals such as MoTe$_2$ and TaS$_2$ which exhibit both coordinations lie on or near the dividing line.

In order to preserve the neutrality of each atom and thereby give the small effective charges characteristic of the Group VI compounds the following bonding scheme has been proposed [360, 372]. Each MX_2 molecule, Figure 47(b), consists of an M atom bound to 6 neighbouring X atoms (each of which is bound to a further two M atoms) giving a total of 18 valence electrons per molecule. The $6M$ electrons form the $d^4 s p$ hybrid orbitals which together with one electron from each X atom form the 6 trigonal prismatic bonds, this accounts for 8 of the valence electrons per molecule. Each X atom contributes a further 2 electrons to the other M bonds leaving 6 valence electrons per molecule unaccounted for. Of these 4 go into non-bonding orbitals on the two chalcogens and it is suggested that the remaining two go to form either an inter or intra-layer bond between two chalcogens.

An earlier bonding scheme proposed [373] that the M atom supplied only 4 bonding electrons, the extra electron(s) of the Group V, VI metal going into a non-bonding d_z^2 metal orbital. While this model is able to account for many of the differences between the Group IV, V, VI compounds it would seem to require the transfer of two chalcogen electrons in order to complete the M—X bonds, this charge transfer is not supported by the infra-red data.

Group IV compounds

As mentioned earlier the Group IV compounds, Table VII, occur either as semiconductors or metals, although the metallic behaviour may, to some extent, be influenced by non-stoichiometry [374].

Transmission measurements, **E**⊥**c**, on cleaved single crystals of ZrS_2 [375], $ZrSe_2$ [376], HfS_2 [375], $HfSe_2$ [375] and TiS_2 [376] grown by iodine vapour transport have indicated that the absorption threshold in these compounds is due to indirect transitions across an energy gap $E_{g'}$ (room temperature) of 1.68, 1.20, 1.96, 1.13 and 0.70 eV respectively. The transmisssion spectra of the Group IV sulphide, selenide crystals (5 K) above threshold [377] are shown in Figure 51 for $\hbar\omega \leq 4.0$ eV and the room temperature reflection spectra in Figure 52 for $\hbar\omega <$ 12 eV. The absorption and reflection spectra all show a broad band (ca. 1.5 eV wide) above threshold which moves to higher energies in the sequence Ti, Zr, Hf, Figure 52(a) and to lower energies in the sequence sulphide, selenide, telluride, Figure 52(b); the higher energy reflectivity peaks behave similarly. The sharp peak E_1 which occurs on the low energy side of the first absorption (reflection) band changes little with temperature between 5 and 77 K and, in ZrS_2, HfS_2,

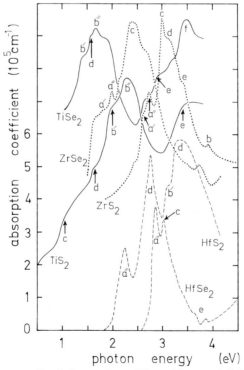

Fig. 51. Transmission spectra (**E**⊥**c**) of some Group IV transition metal dichalcogenide single crystals (5 K) grown by iodine vapour transport [377]. Labelled structure is referred to in the text. For clarity the TiS_2 spectrum has been displaced upwards by 2 units; the ZrS_2, $ZrSe_2$ and $TiSe_2$ spectra by 4 units.

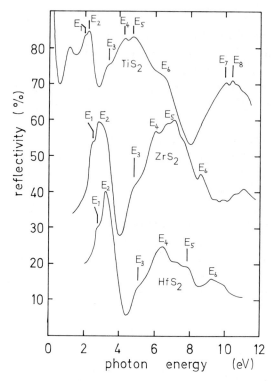

Fig. 52a. Reflectivity spectra ($\mathbf{E} \perp \mathbf{c}$) of TiS$_2$, ZrS$_2$ and HfS$_2$ (room temperature) [375]. For clarity the ZrS$_2$ spectrum has been displaced upwards by 10% and the TiS$_2$ spectrum by 40%.

HfSe$_2$ moves to lower energy under applied hydrostatic pressure [378]. In the Ti compounds the low energy side of the first absorption (reflection) peak is modified by the free carrier absorption which, in TiSe$_2$, TiTe$_2$, Figure 52(b), effectively conceals the absorption threshold.

Transmission measurements on TiS$_2$, ZrS$_2$, HfS$_2$, HfSe$_2$ crystals intercalated with cyclopropylamine [379] have shown a high energy shift of the free carrier absorption in TiS$_2$ and a broadening (in HfSe$_2$) and suppression (in ZrS$_2$, HfS$_2$) of the sharp absorption edge peak. The suppression of the $L_1^+ \to L_1^-$ transition responsible for the absorption edge peak (see later) is attributed to charge transferred from the intercalated molecules to the empty d band of the crystal.

The electron energy loss spectra of the Group IV dichalcogenides show structure due to interband transitions and two prominent plasma resonance peaks due to the 4 non-bonding (π) electrons and the total (16) electrons per molecule. These plasma energies are listed in Table XI. A Kramers–Kronig analysis of the TiS$_2$ ($\mathbf{E} \perp \mathbf{c}$) reflectivity spectrum (80 K) over the range 3 to 10.5 eV [382] gave a derived loss function $-\mathrm{Im}\,\hat{\varepsilon}^{-1}$ peak at 7.3 eV in agreement with the experimental value, Table XI.

Electron energy band structures for the Group IV dichalcogenides have been

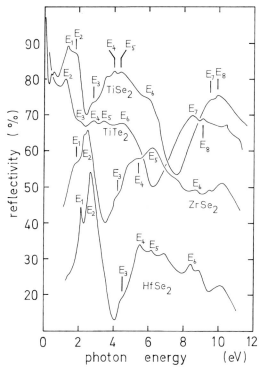

Fig. 52b. Reflectivity spectra ($\mathbf{E} \perp \mathbf{c}$) of TiSe$_2$, TiTe$_2$, ZrSe$_2$ and HfSe$_2$ (room temperature) [375]. For clarity the ZrSe$_2$ spectrum has been displaced upwards by 20%, the TiTe$_2$ spectrum by 30% and the TiSe$_2$ spectrum by 40%.

TABLE XI

Electron energy loss measurements of the plasma energies in the transition metal dichalcogenides. Values are for $\mathbf{q} \perp \mathbf{c}$ unless otherwise stated

Compound	Valence electrons per molecule		Plasma energy (eV)	
	π	Total	π	Total
TiS$_2$ [380]	4	16	6.2 ($\mathbf{q}\|\mathbf{c}$)	21.3 ($\mathbf{q}\|\mathbf{c}$)
			7.3	22.3
TiSe$_2$ [380]	4	16	5.8 ($\mathbf{q}\|\mathbf{c}$)	19.6 ($\mathbf{q}\|\mathbf{c}$)
			6.8	20.7
TiTe$_2$ [380]	4	16	5.8 ($\mathbf{q}\|\mathbf{c}$)	17.2 ($\mathbf{q}\|\mathbf{c}$)
			6.9	18.3
ZrS$_2$ [381]	4	16	3.9	20.3
ZrSe$_2$ [381]	4	16	3.8	19.3
HfS$_2$ [381]	4	16	4.8	21.0
HfSe$_2$ [381]	4	16	3.9	19.4
NbS$_2$ [381]	5	17	8.7	22.3
NbSe$_2$ [381]	5	17	7.7	21
MoS$_2$ [381]	6	18	8.9	23.4
[421]			8.75	23
MoSe$_2$ [381]	6	18	8.0	22.0

calculated by a semiempirical tight binding method [383, 384] which assumes that the threshold is due to indirect transitions, that peak 'a' in Figure 51 is due to a transition at some point along the ML axis and peak 'd' is due to $\Gamma_3^- \to \Gamma_1^+$ transitions. The HfS$_2$ band structure is shown in Figure 53, the HfSe$_2$, ZrS$_2$, ZrSe$_2$ structures are similar. The valence band (v.b) is made up of the chalcogen p levels

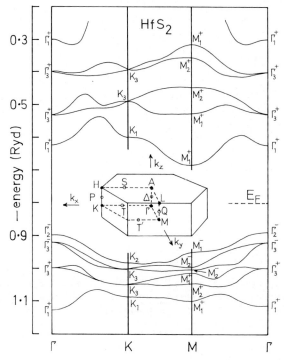

Fig. 53a. An electron energy band structure for HfS$_2$ calculated by a semi empirical tight binding method [383].

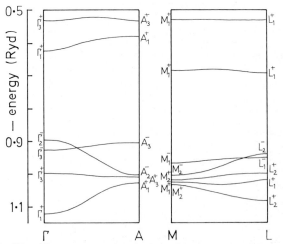

Fig. 53b. Electron energy bands along ΓA and ML for HfS$_2$ [383].

and, at lower energy, the chalcogen s levels while the metal d and s levels form the conduction band (c.b); this is consistent with the ionic picture of these compounds given earlier. The smallest direct gap is at M, between M_1^- and M_1^+, with an indirect gap between Γ_2^- and M_1^+. At Γ the lowest energy allowed transition is $\Gamma_3^- \to \Gamma_1^+$ (fitted to peak 'd'); transitions from the higher Γ_2^- level to Γ_1^+ are forbidden. For $k_z \neq 0$ the direct gap is $L_2^- \to L_1^+$ (fitted to peak 'a'). The remaining labelled structure in Figure 51 is assigned as follows [377]: $-a'' \equiv L_1^- \to L_1^+$, $b' \equiv M_1^- \to M_1^+$, $b'' \equiv M_1^- \to M_1^+$, $c \equiv \Gamma_2^- \to \Gamma_3^+$, $e \equiv \Gamma_3^- \to \Gamma_3^+$, $f \equiv M_1^- \to M_2^+$. The window in the Hf, Zr dichalcogenide spectra around 4 eV, Figure 52, is due, at least in the Hf compounds, to the exhaustion of transitions from the v.b to the Γ_1^+ c.b level.

In going from the Hf to Zr compounds the Γ_3^+ c.b level descends in energy relative to the other levels and in the Ti compounds the Γ_3^+ c.b level occurs below the Γ_1^+ c.b level [384], the $\Gamma_2^- \to \Gamma_3^+$ transition is assumed to be responsible for the additional feature which occurs to the low energy side of E_1 in the Ti compound spectra, Figure 52(a). As in the Zr, Hf compounds the lowest gap was assumed to be the indirect one $\Gamma_2^- \to \Gamma_1^+$, making this zero for $TiTe_2$ (assumed to be a semi metal) also made the $\Gamma_2^- \to \Gamma_3^+$ transition energy zero.

Group V compounds

The Group V dichalcogenides occur as metals and Type II superconductors having superconducting critical temperatures, T_c, of 6.3 K in NbS_2, 7.0 K in $NbSe_2$ and 0.15 K in $TaSe_2$. In $NbSe_2$ at least the value of T_c decreases as the crystal thickness is reduced below ca. 40 Å [385]. The far infra-red transmission spectrum ($\mathbf{E} \perp \mathbf{c}$) of thin superconducting $NbSe_2$ crystals shows a peak at 17.2 ± 0.4 cm^{-1} for a temperature of 1.6 K due to an energy gap of 2.15 meV [386].

The most extensive optical measurements have been on $NbSe_2$ [387–391, 398, 399]. Figure 54(a) shows the $NbSe_2$ reflection spectrum ($\mathbf{E} \perp \mathbf{c}$) over the range 1.7 to 70 eV [387, 388]. The derived energy loss function $-\text{Im}\,\hat{\varepsilon}^{-1}$, Figure 54(a), has a principal peak at 20.9 eV, which is close to the experimental value, Table XI, and two minor peaks at 7.5 and 9.8 eV one of which is close to the observed π plasma energy. The structure which occurs to the high energy side of the second reflectivity threshold at 32 eV has not been positively identified although the minor peak at 56.5 eV could arise from the excitation of Se 3d electrons [388].

Figure 54(b) shows the low energy part of the $NbSe_2$ ($\mathbf{E} \perp \mathbf{c}$) spectrum in more detail [389] together with the $\mathbf{E} \parallel \mathbf{c}$ spectrum obtained from the 'as grown' edge of the plate-like crystal [390, 391]. Also shown in Figure 54(b) are the reflection spectra of 3R NbS_2 ($\mathbf{E} \parallel, \perp \mathbf{c}$) [391], TaS_2, $TaSe_2$ [392]. Transmission spectra of 3R—NbS_2, 2H—$NbSe_2$, 2H—TaS_2 and 2H—$TaSe_2$ [379] are given in Figure 54(c). All these compounds show strong free carrier absorption at low energies which effectively masks the interband absorption threshold. Above threshold the absence of any sharp excitonic structure, as seen in the 2H-Group VI dichalcogenides, is mainly due to the effects of free carrier screening.

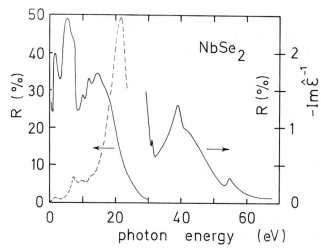

Fig. 54a. The reflectivity spectrum ($E \perp c$) of a NbSe$_2$ crystal (room temperature) over the range 1.7 to 70 eV [387, 388]. The dashed durve is the derived loss function $-\text{Im}\,\hat{\varepsilon}^{-1}$.

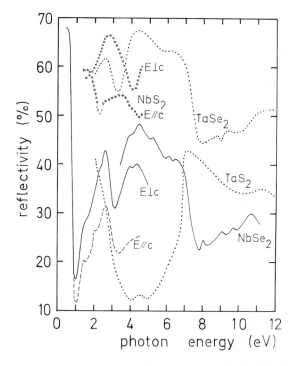

Fig. 54b. The reflection spectra of 2H—NbSe$_2$ (78 K) for $E \perp c$ (full line) and $E \| c$ (dashed line) [389, 391]. The reflection spectra of 3R—NbS$_2$ (78 K) for $E \perp c$ (crosses) and $E \| c$ (circles) [391]. The dotted curves give the reflectivity of TaS$_2$, TaSe$_2$ ($E \perp c$) [392]. For clarity the NbS$_2$ spectra have been displaced upwards by 30% and the TaSe$_2$ spectrum by 20%.

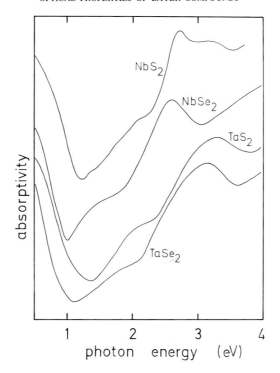

Fig. 54c. The room temperature absorption spectra ($E \perp c$) of 3R—NbS$_2$, 2H—NbSe$_2$, 2H—TaS$_2$ and 2H—TaSe$_2$ [379].

A first principles calculation of the electron energy band structure of 2H—NbSe$_2$ [393] is shown in Figure 55. The twelve lower bands originate from the Se p orbitals while the ten upper bands derive mainly from the metal $4d$ orbitals. In 2H—NbSe$_2$ the Fermi level (dashed line) occurs in the lowest pair of d sub-bands which consist of strongly hybridized combinations of d_z^2 and d_{xy}, $d_{x^2-y^2}$ orbitals [d_z^2 symmetry at Γ; d_{xy}, $d_{x^2-y^2}$ symmetry at K and hybridized combination at M]. The calculated occupied $4d$ bandwidth is 0.5 eV which is close to the photoemission value of ca. 0.7 eV [395]. Similarly the calculated valence p bandwidth is 5.1 eV compared with photoemission values in the range 5 to 7 eV [395, 396]. However the photoemission measurements [395] indicate that the lowest metal d-bands and the Se p bands overlap whereas the calculation predicts a 0.6 eV gap. A similar situation occurs in 2H—MoS$_2$ which has an almost identical band structure [393]; if the $4d$ sub band is moved down to overlap the $3p$ valence band by 0.1–0.2 eV the top of the MoS$_2$ v.b then has d_z^2 symmetry at Γ_4^+ and while this is consistent with paramagnetic resonance studies it complicates the spectral analysis.

The Se s-like core levels occur 14 eV below the 2H—NbSe$_2$ Fermi level and are presumably involved in some of the structure which occurs around this energy in the reflectivity spectrum of Figure 54(a).

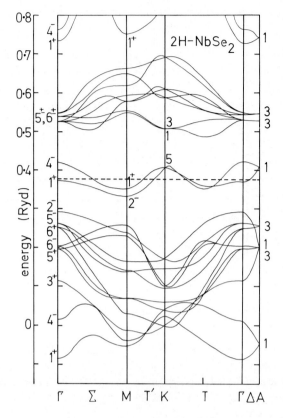

Fig. 55. An electron energy band structure for 2H—NbSe$_2$ calculated by the *APW* method [393].

Group VI dichalcogenides
Some of the first measurements on the optical properties of transition metal dichalcogenides were made on 2H—MoS$_2$ [400–402] which occurs naturally as the mineral molybdenite. Although these, and later, measurements have established many of the macroscopic details of the MoS$_2$ absorption spectrum there are variations in detail which can only be attributed to differing sample purities. Quoted values of the long wavelength ordinary refractive index (**E**⊥**c**), for example, range from 2.65 [400] to 4.33 [403] at room temperature. In view of the ease with which the layer-type dichalcogenides can be intercalated with a wide range of metal atoms [404] the variation in properties of molybdenite should not be surprising.

Figure 56 shows the spectral variation (for $\hbar\omega < 3.8$ eV) of ε_1 and ε_2 (**E**⊥**c**) in single crystals of 2H—MoS$_2$ (290 K) [405], 2H—MoSe$_2$ (77 K) [406], and 2H—MoTe$_2$ (77 K) [407]. The *n*-type MoS$_2$ and *n*-type MoSe$_2$ crystals were grown directly from the powder form while the *n*-type MoTe$_2$ crystals were grown by bromine vapour transport [408]; in the latter case the $\varepsilon_2(\omega)$ spectrum may be

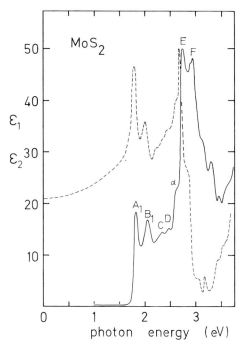

Fig. 56a. The room temperature ε_1 (dashed line) and ε_2 (full line) spectra ($\mathbf{E} \perp \mathbf{c}$) of 2H—MoS$_2$ [405].

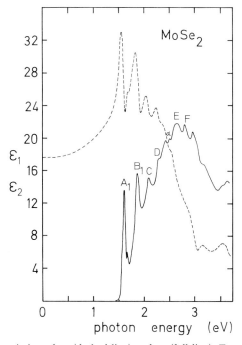

Fig. 56b. The spectral variation of ε_1 (dashed line) and ε_2 (full line), $\mathbf{E} \perp \mathbf{c}$ for 2H—MoSe$_2$ at 77 K [406].

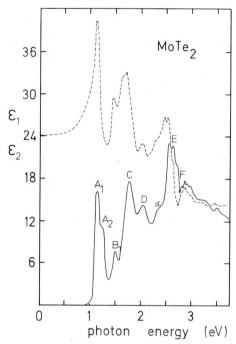

Fig. 56c. The spectral variation of ε_1 (dashed line) and ε_2 (full line), $\mathbf{E} \perp \mathbf{c}$ for 2H—MoTe$_2$ at 77 K [407].

modified by the effect of incorporated Br. The long wavelength ordinary refractive indices of synthetic MoS$_2$, MoSe$_2$, MoTe$_2$ (77 K) are 4.52, 4.22 and 4.9 respectively.

The $\varepsilon_2(\omega)$ spectra in Figure 56 contain a number of strong comparatively narrow absorption bands, labelled A, B, C etc. which broaden with increase in temperature. These band shapes are characteristic of allowed direct transitions into exciton states rather than interband transitions. The identically labelled bands move to lower energies and the $A_1 - B_1$ separation increases as the atomic number of the chalcogen increases.

Figure 57 shows the transmission spectrum ($\mathbf{E} \perp \mathbf{c}$) of a molybdenite crystal (70 K) in the vicinity of absorption bands A_1 and B_1, Figure 56(a), in which the higher quantum number members of each exciton series are visible [409, 410]. The energies at which each member of the A series occurs in synthetic 2H—MoS$_2$ crystals of different thickness are shown in Figure 57(b) [405]. As the crystal thickness (in the $\mathbf{c}$ direction) is reduced the high quantum number (i.e. large radii) excitons are displaced to higher energies to be followed by the smaller quantum number excitons as the thickness is reduced still further. This effect is due to exciton confinement by the crystal surfaces and occurs (by the Uncertainty Principle) when the appropriate exciton diameter becomes comparable with the crystal thickness. Quantitative descriptions of this confinement effect have been derived (a) for a model in which the exciton wave function is made zero for an electron or

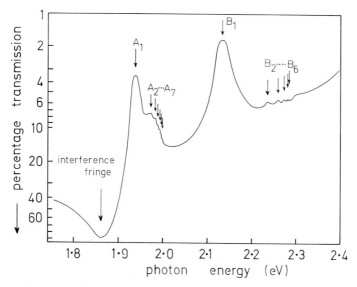

Fig. 57a. The ordinary transmission spectrum of a single crystal of molybdenite (thickness 0.190 μm) at 70 K [409].

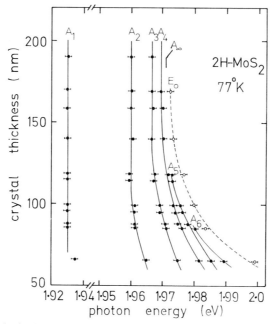

Fig. 57b. Positions (eV) of the A series exciton absorption bands ($E \perp c$) in synthetic 2H—MoS$_2$ crystals (77 K) of different thickness [405]. A_∞ marks the calculated position of the A exciton series limit in thick crystals. The dashed curve passes through the experimental values of E_0 (open circles) the zero field intercept of the Landau oscillation 'fan' diagram, Figure 59.

hole coordinate on or outside the crystal surface [411], Figure 8, and (b) for a model in which the effect of the crystal surfaces is taken into account by means of image charges [412]. Both these models give a partial description of the observed effect. When the crystal thickness $L <$ exciton Bohr radius then the exciton energy $E \propto L^{-2}$, as observed in WSe$_2$ (see later). In molybdenite crystals a few unit cells thick band A_1 weakens and disappears when $L \simeq 3$nm [413].

In 'thick' crystals of synthetic 2H—MoS$_2$ (77 K) the unperturbed energies (eV) of exciton lines A_1—A_4 are given by [405].

$$E_n^A = 1.971 - 0.042/n^2, \quad n = 1, 2, 3, 4$$

these energies agreeing, to within 7 meV, with the corresponding values in molybdenite [409]. This energy relation is the same as (52, 53) which describes three-dimensional delocalized excitons in anisotropic crystals; the two-dimensional exciton equation (54b) does not describe the observed energies of the A or B series excitons in 2H—MoS$_2$. Substituting the experimental value of R_e^A (cm^{-1}) in (52c) gives the exciton radii $r_n^A = 13.82 n^2/\varepsilon$ (nm). For 2H—MoS$_2$ (77 K) $\varepsilon_\perp (= n_0^2) = 20.4$ so that $r_\parallel^A = 0.68 n^2$ (nm). If the birefringence $\Delta (= n_0 - n_e)$ is the same as in high refractive index ($n_0 = 4.33$) molybdenite, viz. $\Delta = 2.3$ [403] then $\varepsilon_\parallel = 4.93$ and $r_\perp^A = 1.4 n^2$ (nm). These A series exciton radii in synthetic 2H—MoS$_2$ are smaller than in molybdenite ($\varepsilon_\perp = 6.76$, $\varepsilon_\parallel = 2.74$ [401]) and in agreement with this the onset of exciton confinement in 2H—MoS$_2$, Figure 57(b), occurs at smaller crystal thicknesses than in molybdenite [409]. The ratio of exciton radius to crystal thickness at the onset of confinement is approximately the same for both natural and synthetic crystals.

If scaling condition (52a) is satisfied then the experimental values of R_e^A, $\varepsilon_\perp$, $\varepsilon_\parallel$ in 2H—MoS$_2$ (77 K) when substituted in (52b) gives $\mu_\parallel^A = 1.28 m_0$ and $\mu_\perp^A = 0.31 m_0$.

In a similar manner the B exciton series in synthetic 2H—MoS$_2$ show the effects of confinement. The thick crystal energies (eV) of B_1—B_4 (77 K) are given by $E_n^B = 2.267 - 0.134/n^2$, $n = 1, 2, 3, 4$. Hence $r_\parallel^B = 0.26 n^2$ (nm) and $r_\perp^B = 0.53 n^2$ (nm) while $\mu_\parallel^B = 4.10 m_0$ and $\mu_\perp^B = 0.99 m_0$.

In the rhombohedral polytype 3R—MoS$_2$, Figure 48(b), the exciton bands occur at a lower energy [414], the shift being greatest for the B_1 exciton band. The $A_1 B_1$ separation in the 3R polytype is about three quarters that in the 2H - polytype which again shows that layer–layer interactions are important.

Over the temperature range 350 to 4 K A_1, A_2 show an identical energy shift to higher energies [405], so that R_e^A can be considered temperature independent, although earlier measurements did show a dependence [415]. Between 350 and 150 K the constant temperature coefficients $(\partial E/\partial T)p$ of A_1, A_2 and B_1 are -0.40 meV K^{-1} and -0.47 meV K^{-1} respectively; the coefficients then steadily decrease with temperature and become immeasurably small near 4 K. A transition energy E is a function of the temperature, pressure and volume of the solid so that [378]

$$\left(\frac{\partial E}{\partial T}\right)_p = \left(\frac{\partial E}{\partial T}\right)_v - \left[\frac{\alpha_\parallel + 2\alpha_\perp}{\beta_\parallel + 2\beta_\perp}\right]\left(\frac{\partial E}{\partial p}\right)_T,$$

where α $(= 1/V \cdot (\partial V/\partial T)_p)$ and β $(= -1/V(\partial V/\partial p)_T)$ are the expansion coefficient and isothermal compressibility parallel ($\parallel$) and perpendicular ($\perp$) to the crystal c axis. The first term on the right hand side of the equation expresses the phonon dependence of the energy levels and the second term the energy change due to lattice expansion. The pressure coefficients $(\partial E/\partial p)_T$ of A_1 and B_1 are 1.4 and 1.6 10^{-6} eV bar^{-1} respectively in 2H—MoS$_2$ (80 K) and 0.9, 1.2 10^{-6} eV bar^{-1} in the 3R polytype [378], these values increase with temperature. Using appropriate values of α and measured values of β, $(\partial E/\partial p)_T$ the lattice dilation term becomes approximately -1.0×10^{-5} eV K^{-1} at room temperature. Since $(\partial E/\partial T)_p = -4 \times 10^{-4}$ eV K^{-1} for A_1 so the electron–lattice term $(\partial E/\partial T)_V \simeq -4 \times 10^{-4}$ $eV K^{-1}$ accounts for the temperature shift of A_1.

The transverse electroreflectance ($\boldsymbol{\xi} \perp \mathbf{c}$, $\mathbf{E} \perp \mathbf{c}$) spectrum of MoS$_2$ (50 K) from the absorption threshold to 4.0 eV [416] is shown in Figure 58. The electroreflectance (e.r) spectrum shows a lot of structure in the region of A, B as does the longitudinal electroabsorption ($\boldsymbol{\xi} \parallel \mathbf{c}$, $\mathbf{E} \perp \mathbf{c}$) spectrum [417] shown in the inset to Figure 58. The absorption and e.r spectrum both indicate that C is composed of two overlapping absorption bands. In the region of D the e.r signal is small but a strong signal occurs at 3.6 eV corresponding to absorption band α. In the electroabsorption spectrum, Figure 58, peaks a, b, c are attributed to a broadened

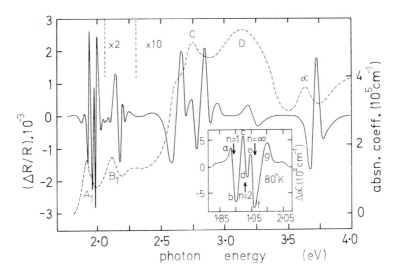

Fig. 58. The transverse electroreflectance spectrum of MoS$_2$ at 50 K [416]. The dashed line represents the molybdenite absorption spectrum at 77 K [401]. Inset is the electroabsorption spectrum (longitudinal configuration, $\boldsymbol{\xi} \parallel \mathbf{c}$) of an MoS$_2$ crystal $\sim$600 Å thick at 80 K [417].

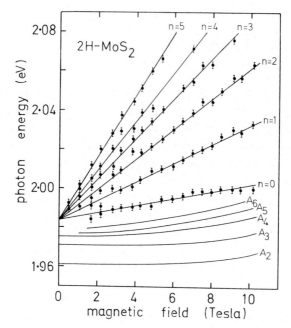

Fig. 59. Diamagnetic shift of the A series exciton absorption lines and positions (eV) of the magneto absorption peaks for various values of the magnetic field **H**∥**c** in a synthetic 2H—MoS$_2$ crystal (77 K) of thickness 88 nm [405].

A_1 exciton line, which is also shifted to lower energy; peaks c, d, e to a field broadened A_2 line and strong peaks f, g to the continuum absorption.

In an applied magnetic field the A and B series exciton lines show a diamagnetic shift [418, 409], lines A_1 and B_1 show Zeeman splitting [405] and the continuum absorption on the high energy sides of the A and B series excitons shows magneto oscillations due to Landau level transitions [409, 405]. Some of these properties are contained in Figure 59 which shows the diamagnetic shift of the A series exciton lines and energies of the magneto absorption maxima at different field strengths **H**∥**c**. The diamagnetic energy shift ΔE_D of the exciton lines is given by $\Delta E_D = S_\parallel H^2$ where $S_\parallel$ decreases with relevant exciton confinement. Thick crystal values of $S_\parallel$ yield $\mu_\perp$, the exciton reduced mass, (131). The common intercept E_0 of the fan lines through the magneto absorption peaks is taken as the exciton series limit, (142); this also is crystal thickness dependent as shown in Figure 57(b). The reduced mass appropriate to the magneto oscillations is much smaller than the exciton reduced mass deduced from the Rydberg constant or diamagnetic shift; this has not yet been explained.

The reflection spectrum (**E**⊥**c**) of molybdenite over the range 1.7 to 70 eV [387, 388] is given in Figure 60; these measurements agree with others made over the range 1 to 12.5 eV [389, 419]. The derived loss function $-\text{Im}\,\hat{\varepsilon}^{-1}$ shows a small peak at 8.8 eV and a large peak at 23.3 eV in good agreement with electron energy loss measurements, Table XI. The threshold near 37 eV is identified with

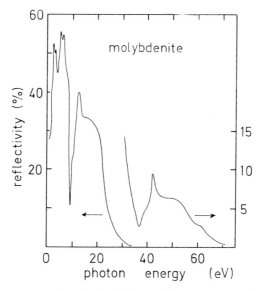

Fig. 60. The reflection spectrum ($\mathbf{E}\perp\mathbf{c}$) of molybdenite (room temperature) over the range 1.7 to 70 eV [387, 388].

the excitation of Mo 4p electrons into d and s conduction states (peak at 43 eV). The shoulder at 60 eV agrees with the excitation energy of 4s electrons of molybdenum which occur in the metal at 61.8 eV [422].

The $\varepsilon_2(\omega)$ spectrum of 2H—MoSe$_2$ crystals grown directly from the powder is shown in Figure 56(b); crystals grown by bromine transport occur as mixed 2H and 3R polytypes [406]. At the absorption edge exciton lines A_1, A_2 are observed whose energies give (subject to the conditions of (52, 53)) an energy gap E_g^A (77 K) = 1.69 eV. The energy separation of A_1, B, is 0.263 eV. At higher energies the reflectivity spectrum of vapour transport 2H—MoSe$_2$ is as shown in Figure 61 [423]. The reflectivity minimum at 8.43 eV is characteristic (c.f MoS$_2$) of a plasma resonance, Table XI, and marks the exhaustion of one set of interband transitions.

The $\varepsilon_2(\omega)$ spectrum of vapour grown 2H—MoTe$_2$ crystals, Figure 56(c), contains what seem to be line pairs A_1, B_1; C, D; E, F which, by their similarity to A and B, are evidently due to excitonic transitions. The shoulder representing A_2 is not sufficiently well resolved to enable E_g^A to be calculated. The $A_1 B_1$ separation is 0.348 eV (77 K), this doublet disappears in the $\mathbf{E}\|\mathbf{c}$ spectrum [391]. Some of the lines, such as E, show a splitting not observed in 2H—MoSe$_2$, which may be real or the result of stacking polytypism since MoS$_2$ and MoSe$_2$ both show a shift in position of the higher energy bands in going from the 2H to 3R polytype. Over the range 4 to 14 eV, Figure 61, the 2H—MoTe$_2$ reflection spectrum resembles that of 2H—MoS$_2$, MoSe$_2$. The reflectivity minimum (corresponding to the plasma energy) occurs at 7.46 eV (85 K) whereas the derived loss function $-\text{Im}\,\hat{\varepsilon}^{-1}$ peaks at 6 eV at room temperature [424].

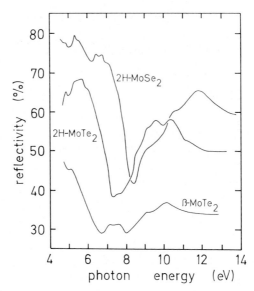

Fig. 61. The **E**⊥**c** reflectivity spectra over the range 4.5 to 14 eV of 2H—MoSe$_2$, 2H(α)—MoTe$_2$ and β—MoTe$_2$ at 85 K [423]. For clarity the α—MoTe$_2$ curve has been displaced vertically by 20% and the 2H—MoSe$_2$ curve by 30%.

The β form of MoTe$_2$ has a distorted layer structure resulting in Mo—Mo pairing. The β—MoTe$_2$ reflectivity spectrum, Figure 61, resembles that of the tungsten, rather than molybdenum, dichalcogenides and this may be because in the 2H tungsten compounds the nearest neighbour metal–metal overlap is greater than in the corresponding (smaller atom) molybdenum compounds.

For all three 2H molybdenum compounds of Figures 56, 60, 61, the structure present in the reflectivity spectra shows a systematic shift to lower energies in the sequence MoS$_2$, MoSe$_2$, MoTe$_2$ which indicates that, as expected, the band structures of these compounds are closely related.

Electron energy band structures for 2H—MoS$_2$ and related compounds have been obtained using a semi-empirical tight binding method [425], a LCAO method [426], a molecular orbital approach [427], an APW method [393] and a linear combination of muffin tin orbitals (LCMTO) method [428]. Any 2H—MoS$_2$ band structure has to (1) satisfy the photoemission results [395, 396, 429], c.f NbSe$_2$, that (a) the intrinsic energy gap is at least 1 eV wide, (b) the metal d-band (c.a 1.5 eV wide) overlaps the chalcogen p-band by 0.1 to 0.2 eV (2) satisfy the conclusions of electron paramagnetic resonance measurements on Nb, As acceptor levels in MoS$_2$ [430] that the upper v.b has d_z^2 character. On this basis the APW [393] and LCMTO [428] calculations seem most promising. The APW calculation gives a 2H—MoS$_2$ band structure which resembles that of 2H—NbSe$_2$, Figure 55, except that the bands are depressed by about 0.1 Ryd. leaving the Fermi level in the band gap; the limitations of this band structure are those for NbSe$_2$.

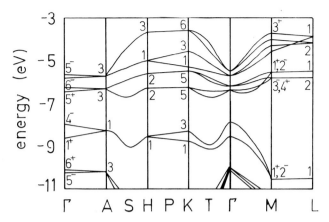

Fig. 62. An electron energy band structure for 2H—MoS$_2$ calculated by a linear combination of muffin-tin orbitals method [428].

The LCMTO band structure is given in Figure 62. The S p-bands extend from -10.2 to -16.2 eV. The d_z^2 band extends from -8.0 to -10.7 and thus overlaps the S p band by 0.5 eV as required. The remaining d bands mix strongly with the S p states (to form d/p bands) while the Mo $5s$ state lies above and separate from the d/p bands. The calculated energy gap of 1.1 eV is indirect and there is some agreement between the predicted and observed absorption features, Figure 56(a), when identified as in Table XII. On this scheme the A, B exciton pair is due to spin orbit (s.o) splitting (~ 0.2 eV) of the A_3 c.b level since A_1 level does not s.o split. In the tight binding calculation [425] the A, B doublet is attributed to Γ point transitions which do not feature in Table XII. The C, D line pair is assigned to exciton states of the $H_1 \rightarrow H_2$ transition, the increase in intensity of these lines in the sequence MoS$_2$, MoSe$_2$, Figure 56, is attributed to the increased p content of the d/p bands with increasing anion size.

As yet there is no general agreement on the nature of the transitions giving rise to the structure in the Group VI dichalcogenide spectra and detailed joint density of states calculations are required to settle this matter.

TABLE XII

Assignment of theoretical transitions to observed optical features ($\mathbf{E} \perp \mathbf{c}$) in 2H—MoS$_2$ [428]; refer to Figure 62

Observed feature (room temperature), Figure 62	Assigned transition energy
A ($E_g^A = 1.89$ eV)	$A_1 \rightarrow A_3$ (2.0 eV)
B ($E_g^B = 2.22$ eV)	
C (2.34 eV)	$H_1 \rightarrow H_2$ (2.39 eV)
D (2.47 eV)	
E (2.72 eV)	$A_1 \rightarrow A_3$ (2.58 eV)
F (2.9 eV)	$H_1 \rightarrow H_2$ (3.17 eV)

Tungsten dichalcogenides

The reflectivity spectra (78 K) of single crystals of 3R—WS$_2$ and 2H—WSe$_2$ (grown by Br vapour transport) are shown in Figure 63 [391]; the transmission spectra of the iodine grown compounds have also been measured [377]. In both compounds the **E**⊥**c** spectra show sharp exciton peaks above threshold with further strong peaks at higher energies attributed to interband transitions. In terms of the two dimensional WS$_2$ band structure shown in Figure 64 [425] the labelled **E**⊥**c** reflectivity structure is identified as follows [391]: – A, B excitons with $\Gamma_3^- \to \Gamma_3^-$, $d = \Gamma_1^+ \to \Gamma_3^+$, $C = Q_2^- \to Q_2^-$, $D = P_3^- \to P_3^-$. For **E**∥**c** transitions between the highest v.b and lowest c.b are only weakly allowed due to the d-type character of these bands; the lowest energy allowed transitions, **E**∥**c**, are $Q_2^+ \to Q_2^-$ giving the transitions at 3.02 eV in WS$_2$, 2.89 eV in WSe$_2$. Transitions $\Gamma_3^+ \to \Gamma_3^-$ and $\Gamma_3^- \to \Gamma_3^+$, also allowed for **E**∥**c**, are thought to be responsible for the feature at 3.42 eV in the WS$_2$ (**E**∥**c**) spectrum; in WSe$_2$ they occur near 2.5 eV.

The ε_1, ε_2 spectra (**E**⊥**c**) of p-type 2H—WSe$_2$ (77 K) crystals grown directly from the powder are shown in Figure 65. Interference fringe measurements give the long wavelength ordinary refractive index n_0 (room temperature) as 4.35 and

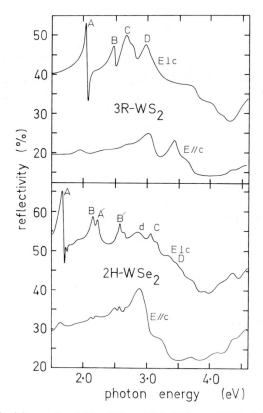

Fig. 63. The reflectivity spectra of 3R—WS$_2$ and 2H—WSe$_2$ at 78 K for **E**∥**c** and **E**⊥**c** [391].

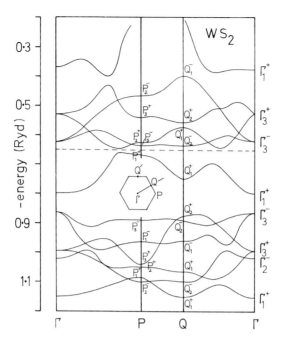

Fig. 64. A two-dimensional energy band structure for WS_2 obtained by a semi-empirical tight binding method [425]. The dashed line shows the position of the Fermi energy and the inset shows the two-dimensional Brillouin zone.

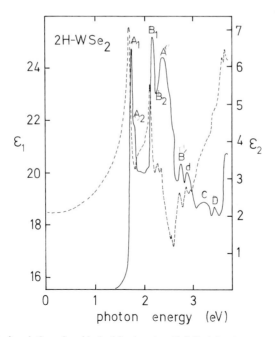

Fig. 65. The spectral variation of ε_1 (dashed line) and ε_2 (full line) for 2H—WSe_2 (77 K) $\mathbf{E}\perp\mathbf{c}$ [407].

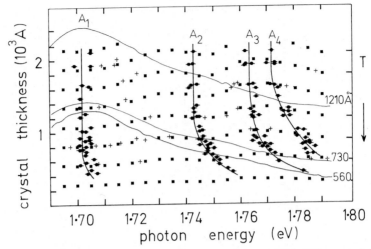

Fig. 66. The transmission spectra, in the A series exciton region, of 2H—WSe$_2$ crystals (117 K) of thickness 0.121, 0.073 and 0.056 μm respectively. Also shown are the measured positions —●— of the A series absorption bands in crystals of different thickness and the observed + and calculated ■ positions of the transmitted interference minima [394].

this rises to ca. 6.9 at A_1 (117 K), ellipsometric measurements give $n_0 = 5$ at A_1 (room temperature) [431].

Figure 66 shows the tranmission spectra ($\mathbf{E} \perp \mathbf{c}$), in vicinity of A_1, of 2H—WSe$_2$ crystals of different thickness [394]. With decreasing crystal thickness L the A series exciton peaks are displaced to higher energies as a result of exciton confinement, c.f MoS$_2$, and in very thin crystals, Figure 67, the $n = 1$ (A_1) exciton shows [432] the expected L^{-2} energy dependence.

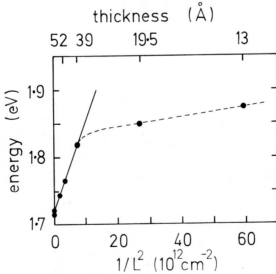

Fig. 67. Energy of exciton peak A_1 plotted against $1/L^2$, where L is the crystal thickness of WSe$_2$ crystals on an Epoxy substrate at 77 K [432].

When the crystal thickness $L > 200$ nm the exciton energies are independent of L; in this case the energies (eV) of absorption bands A_n ($n = 2, 3, 4, 5$) at 117 K are given by [394] $E_n = 1.782 - 0.158/n^2$. Band A_1 (at 1.702 eV) occurs to the high energy side of the position predicted by this three-dimensional equation, (52). For the $n = 2, 3, 4, 5$ A series excitons the radii parallel to the c axis are given by (52o), $r_\parallel^A = 0.245 n^2$ (nm). Since WSe$_2$ is uniaxial negative so $r_\perp^A > r_\parallel^A$. Similarly the exciton reduced mass $\mu_\parallel = 4.05 m_0$ and $\mu_\perp < \mu_\parallel$.

Thermoreflectance [431] and electroreflectance [434] measurements on WSe$_2$ have only detected A_1 and A_2 which, see above, give an incorrect value for R_e^A. The hydrostatic pressure coefficients of A_1 and B_1 are positive whereas peaks A', B' are both moved to lower energies with increasing pressure [378].

10.9. Graphite

The crystal structure of hexagonal graphite consists of planes (layers) of C atoms widely spaced parallel to each other along the c axis; the stacking sequence of the layers is AB, AB, ... Natural graphite can be a mixture of hexagonal and rhombohedral structures. In the rhombohedral component, which disappears on heat treatment, the layers have the stacking sequence ABC, ABC [441]. Single crystals of rhombohedral graphite have not been prepared so there is no information about its optical properties although the band structure has been calculated [442, 443].

One of the calculated [437–440] band structures of hex. graphite is given in Figure 68. In the single layer (2D) model, Figure 68(a) the Fermi energy level E_F passes through state P_3^-, characterizing the material as a semiconductor with vanishing energy gap. In the three dimensional (3D) model, which includes the weak interlayer interaction, the 2D energy bands are slightly split (<0.15 eV) over most of the B.Z but at certain **k** the splitting is larger and important. For example states at P_3^- are split into P_2^-, P_3^-, P_1^- and the energy band dependence on $\mathbf{k}_z$, e.g. along HPH, Figure 68(b) determines the complex Fermi surface and semi metallic properties of graphite. Photoemission measurements give the splitting at P and Q as 0.8 ± 0.1 eV [436].

The spectral variation of the measured [444] normal incidence reflectivity (unpolarized light, $\mathbf{E} \perp \mathbf{c}$) $R_\perp$ of a freshly cleaved sample is shown in Figure 69, together with $\varepsilon_{1\perp}$, $\varepsilon_{2\perp}$ over the range 0.006 to 0.13 eV found from reflection measurements on pyrolytic graphite [445]. Also shown in Figure 69 is the reflectivity $R_\parallel(\mathbf{E} \parallel \mathbf{c})$ over the range 2 to 10 eV and the variation of $n_\perp$, $\varkappa_\perp$, $n_\parallel$, $\varkappa_\parallel$ over the range 2 to 5 eV as determined [446] by a Fresnel analysis of the light (polarized perpendicular, parallel, to the plane of incidence) reflected at different angles of incidence. Values for $(n, \varkappa)_{\parallel,\perp}$ have also been obtained [447] by measuring the normal incidence reflectivity of a $10\bar{1}2$ graphite face; when $\mathbf{E} \perp \mathbf{c}$ the reflectivity is that of the basal plane but when **E** is rotated through $\pi/2$ then measured n, $\varkappa$ are functions of $n_\perp$, $\varkappa_\perp$, $n_\parallel$, $\varkappa_\parallel$ and δ the angle that the wave normal makes with **c**. Hence knowing $n_\perp$, $\varkappa_\perp$, δ so $n_\parallel$, $\varkappa_\parallel$ are determined. Both sets of

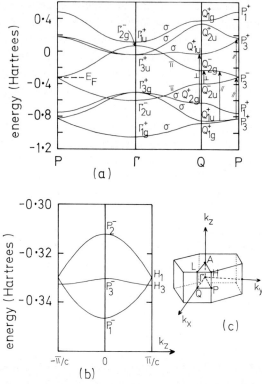

Fig. 68. (a) Energy band structure for graphite in the single layer crystal model. [438]. Allowed transitions are indicated for polarizations parallel ($\parallel$) and perpendicular ($\perp$) to the c axis. (b) Energy band dependence upon k_z along the edge HPH of the three-dimensional Brillouin zone for multiple-layer graphite lattice. Here the c.b and v.b overlap so that graphite behaves like a semi-metal [438]. (c) Brillouin zone for graphite. For the single layer crystal the Brillouin zone is the hexagon formed by the intersection with the plane $k_z = 0$.

measurements [446, 447] show that $n_\parallel$, $\varkappa_\parallel$ are constant, with $n = 1.54$ [446] (1.81) [447], $\varkappa = 0$ over the range 2 to 5 eV, Figure 69. The spectral variation of $\varepsilon_{2\parallel}$ over the range 6 to 30 eV, Figure 69, has been determined [448] from electron energy loss measurements [449]. Structure observed in the thermoreflectance spectrum [435] could be due not only to a shift and broadening of interband transitions, but also to the broadening of the step in the Fermi distribution.

Over the energy range 0.006 to 0.13 eV $\varepsilon_{2\perp}$ shows peaks at [445] 0.008 eV (150 μm), 0.11 eV (110 μm) and 0.02 eV (60 μm), it is evident from Figure 68(a) that such small energy differences between the bands occur only at the B.Z edge, consequently an explanation for these infra-red bands has been sought [450] in terms of the graphite band structure along HKH, including s.o splitting, Figure 68(c). On this model two of the peaks can be attributed to transitions between the degenerate (s.o split) bands near the points Q and K respectively but the third peak is unexplained on this scheme. The weak reflectance structure near 0.8 eV is attributed [438] to $P_2^- - P_1^-$, Figure 68(b).

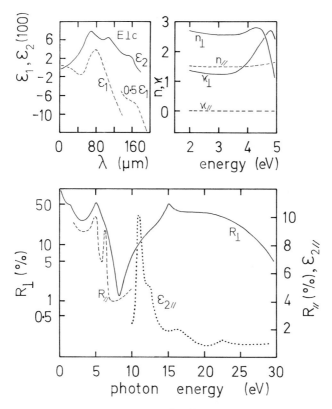

Fig. 69. The spectral variation of $R_\perp$ for graphite [444] (full line), $R_\parallel$ for polished pyrolytic graphite [446] (dashed curve), $\varepsilon_{2\parallel}$ for graphite determined from the electron energy loss spectrum [448] (dotted curve). Also shown is ε_1, ε_2 ($\mathbf{E}\perp\mathbf{c}$) over the range 10 to 190 μm for pyrolytic graphite [445] and $(n, \varkappa)_{\parallel,\perp}$ between 2 and 5 eV for graphite [446]. Subscripts $_{\parallel,\perp}$ refer to $\mathbf{E}\parallel\mathbf{c}$ and $\mathbf{E}\perp\mathbf{c}$ respectively.

At higher energies the $\varepsilon_{2\perp}$ spectrum, obtained from Kramers-Krönig analysis of $R_\perp$, Figure 69, shows peaks [444, 446] at ca. 4.5 and 14.5 eV, the 4.5 eV peak having [446] a shoulder at ca. 6 eV. Band structure calculations [446, 438] attribute the 4.5 eV peak to $Q_{2g}^- - Q_{2u}^-$ (allowed for $\mathbf{E}\perp\mathbf{c}$, Figure 68(a)); the 6 eV shoulder to $\Gamma_{3g}^+ - \Gamma_{3u}^+$ [446] (allowed $\mathbf{E}\perp\mathbf{c}$) or structure in the π bands near Q [438]. Figure 68(a) gives $\Gamma_{3g}^+ - \Gamma_{3u}^+$ transitions at ca. 12.2 eV followed by transitions between the highest occupied σ band and the first σ c.b at (the peak) 14 eV, then $Q_{2g}^+ - Q_{1u}^+$ transitions at ca. 16.3 eV. Other calculations [446] attribute the 14.5 eV peak to $Q_{2u}^+ - Q_{2g}^+$.

The measured variation of $\varepsilon_{1\perp}$, $\varepsilon_{2\perp}$, over the range 0 to 30 eV can be well described [452] by a two oscillator equation which considers $N_1(\pi)$ electron oscillators (cm^{-3}) having resonant frequency $\hbar\omega_1 = 4$ eV and $N_2(\sigma)$ electron oscillators (cm^{-3}) of resonant frequency $\hbar\omega = 14$ eV. These classical equations for $\varepsilon_{1\perp}$, $\varepsilon_{2\perp}$ predict that $\text{Im}(\hat{\varepsilon})^{-1}$ has peaks at 7.5 and 28 eV compared with the corresponding measured electron energy loss peaks at [449] 7.2 and 27 eV.

The $\mathbf{E}\parallel\mathbf{c}$ spectrum, Figure 69 shows an additional peak near 6 eV which is evidently allowed for $\mathbf{E}\|\mathbf{c}$ but forbidden for $\mathbf{E}\perp\mathbf{c}$. This peak is attributed [446] to transitions between the highest σ v.b and lowest π c.b. The sharp peak ($\mathbf{E}\|\mathbf{c}$) at 11 eV is identified [448, 438] with transitions between the almost parallel bands $\sigma_2 \to \pi_2$ and the broad peak at 16 eV with π_1 (v.b) $\to \sigma$ (c.b).

Using graphs of n_{eff}, the effective number of electrons, versus ω obtained from measured $\varepsilon_2(\omega)$ [448], it is found that $(n_{\text{eff}})_\perp$ saturates at 4 electrons per atom by $\hbar\omega = 25$ eV whereas $(n_{\text{eff}})_\|$ is still far from saturation at $\hbar\omega = 30$ eV. The difference between these two cases is a consequence of the selection rules. For $\mathbf{E}\perp\mathbf{c}$ at the point Q, Figure 68(a) there are allowed transitions to the four lowest c.b from all the occupied bands whereas for $\mathbf{E}\|\mathbf{c}$ no such transitions are possible from Q_{1g}^+, Q_{2g}^+ points [451] so that $(n_{\text{eff}})_\|$ is still far from saturation at 30 eV.

References

1. B. L. Evans and P. A. Young: *Proc. Roy. Soc. A* **284** (1965), 402.
2. B. L. Evans and P. A. Young: *Proc. Roy. Soc. A* **298** (1967), 74.
3. See, for example, L. I. Schiff, *Quantum Mechanics*, McGraw-Hill, New York, 1968.
4. L. van Hove: *Phys. Rev.* **89** (1953), 1189.
5. J. C. Phillips: *Solid State Physics* **18**, Academic Press, New York, 1966.
6. N. J. Doran and P. G. Harper: *Phys. Stat. Sol.* **48** (1971), 223.
7. See, for example, R. A. Smith: *Wave Mechanics of Crystalline Solids*, Chapman and Hall, London, 1961.
8. R. L. Hartman: *Phys. Rev.* **127** (1962), 765.
9. J. L. Brebner: *J. Phys. Chem. Solids* **25** (1964), 1427.
10. Review article by D. S. McClure: *Solid State Phys.* **8** (1959), 1.
11. Review article by H. C. Wolf: *Solid State Phys.* **9** (1959), 1.
12. G. F. Koster and J. C. Slater: *Phys. Rev.* **95** (1954), 1167.
13. G. H. Wannier: *Phys. Rev.* **52** (1937), 191.
14. N. F. Mott: *Trans. Faraday Soc.* **34** (1938), 500.
15. H. Haken: *J. Phys. Chem. Solids* **8** (1959), 166.
16. J. Friedel: *Phil. Mag.* **43** (1952), 153.
17. R. C. Casella: *J. Appl. Phys. (USA)* **34** (1963), 1703.
18. G. Dresselhaus: *Phys. Rev.* **106** (1957), 76.
19. J. J. Hopfield and D. G. Thomas: *Phys. Rev.* **122** (1961), 35.
20. P. G. Harper and J. A. Hilder: *Phys. Stat. Sol.* **26** (1968), 69.
21. H. I. Ralph: *Solid State Commun.* **3** (1965), 303.
22. M. Shinada and S. Sugano: *J. Phys. Soc. Japan* **21** (1966), 1936.
23. H. I. Ralph: *J. Phys. (France)* **28** (1967), 57.
24. J. L. Brebner, R. Fivaz, and E. Mooser: *Nature* **210** (1966), 931.
25. R. A. Satten and S. Nikitine: *Phys. Kondens Materie* **1** (1963), 394.
26. B. L. Evans and P. A. Young: *Phys. Stat. Sol.* **25** (1968), 417.
27. G. Jones: *Phys. Rev.* **B4** (1971), 2069.
28. R. J. Elliott: *Phys. Rev.* **108** (1957), 1384.
29. R. S. Knox: *Solid State Phys., Suppl.* **5** (1963).
30. J. C. Phillips: *Solid State Phys.* **18** (1966), 55.
31. C. B. Duke and B. Segal: *Phys. Rev. Letters* **17** (1966), 19.
32. S. Flugge and H. Marschall: *Rechenmethoden der Quantentheorie*, Springer-Verlag, 1952.
33. B. Velicky and J. Sak: *Phys. Stat. Sol.* **16** (1966), 147.
34. Y. Toyozawa, M. Inoue, T. Inui, M. Okazaki, and E. Hanamura: *J. Phys. Soc. Japan* **21** (1966), Suppl. 133.
35. As [34] p. 1850.

36. J. Hermanson: *Phys. Rev.* **166** (1968), 893.
37. Y. Toyozawa: *J. Phys. Chem. Solids* **8** (1959), 289.
38. Y. Toyozawa: *Progr. Theoret. Phys.* **20** (1958), 53.
39. Y. Toyozawa: *Progr. Theoret. Phys.* **27** (1962), 89.
40. Y. Toyozawa: *J. Phys. Chem. Solids* **25** (1964), 59.
41. P. Drude: *Theory of Optics*, Longmans Green and Co., New York 1902.
42. C. Zener: *Nature* **132** (1933), 968.
43. R. de L. Kronig: *Nature* **133** (1934), 211.
44. W. P. Dumke: *Phys. Rev.* **124** (1961), 1813.
45. H. Y. Fan, W. Spitzer, and R. J. Collins: *Phys. Rev.* **101** (1956), 566.
46. R. Rosenberg and M. Lax: *Phys. Rev.* **112** (1958), 843.
47. C. Herring and E. Vogt: *Phys. Rev.* **101** (1956), 944.
48. G. N. Watson: *A Treatise on the Theory of Bessel Functions*, Macmillan and Co., New York 1945.
49. K. Tharmalingam: *Phys. Rev.* **130** (1963), 2204.
50. D. E. Aspnes: *Phys. Rev.* **147** (1966), 554; also **153** (1967), 972.
51. L. Fritsche: *Phys. Stat. Sol.* **13** (1966), 487.
52. W. V. Houston: *Phys. Rev.* **57** (1940), 184.
53. W. Franz: *Z. Naturforsch.* **13** (1958), 484.
54. L. V. Keldysh: *Sov. Phys. JETP* **7** (1958), 788.
55. M. Chester and L. Fritsche: *Phys. Rev.* **139** (1965), A518.
56. Y. Yacoby: *Phys. Rev.* **140** (1965), A263.
57. J. Callaway: *Phys. Rev.* **130** (1963), 549; also **134** (1964), A998; also **143** (1966), 564.
58. E. N. Adams: *Phys. Rev.* **107** (1957), 698 and earlier references therein.
59. D. E. Aspnes, P. Handler, and D. F. Blossey: *Phys. Rev.* **166** (1968), 921.
60. P. W. Argyres: *Phys. Rev.* **126** (1962), 1386.
61. H. A. Antonsiewicz: *Handbook of Mathematical Functions*, (ed. by M. Abramowitz and I. A. Stegun), U.S. Dept. of Commerce, National Bureau of Standards, Washington, D.C. 1964.
62. I. P. Batra: *J. Phys. Soc. Japan* **26** (1969), 1419.
63. F. Aymerich and F. Bassani: *Nuovo Cimento* **48B** (1967), 358.
64. J. C. Phillips: *Phys. Rev.* **146** (1966), 584.
65. R. Enderlein and R. Keiper: *Phys. Stat. Sol.* **19** (1967), 673. Also **23**, 127.
66. B. O. Seraphin and N. Bottka: *Phys. Rev.* **145** (1966), 628.
67. D. E. Aspnes: referred to in [68].
68. M. Cardona: *Solid State Phys., Suppl.* **11** (1969).
69. C. M. Penchina: *Phys. Rev.* **138** (1965), A924.
70. L. Fritsche and M. Chester: *Phys. Rev.* **139** (1965), A518.
71. L. Fritsche: *Phys. Stat. Sol.* **11** (1965), 381.
72. D. F. Blossey: *Phys. Rev.* **B2** (1970), 3976; also **3** (1971), 1382.
73. E. Yang: *Phys. Rev.* **B4** (1971), 2046.
74. H. I. Ralph: *J. Phys. C. Proc. Phys. Soc.* **1** (1968), 378.
75. J. D. Dow and D. Redfield: *Phys. Rev.* **B1** (1970), 3358.
76. J. E. Rowe and D. E. Aspnes: *Phys. Rev. Letters* **25** (1970), 162.
77. N. Bottka and J. E. Fischer: *Phys. Rev.* **B3** (1971), 2514.
78. J. M. Luttinger and W. Kohn: *Phys. Rev.* **97** (1955), 869; and **98**, 915.
79. J. M. Luttinger: *Phys. Rev.* **102** (1956), 1030.
80. A. Baldereschi and F. Bassani: *Phys. Rev. Letters* **19** (1967), 66.
81. J. G. Mavroides: *Optical Properties of Solids*, North Holland Publ. Co., Amsterdam 1972, 406.
82. R. J. Elliott and R. Loudon: *J. Phys. Chem. Solids* **15** (1960), 196.
83. G. Dresselhaus: *J. Phys. Chem. Solids* **1** (1956), 14.
84. B. Lax and S. Zwerdling: *Prog. Semicond.* **5** (1960), 221.
85. B. L. Evans and P. A. Young: *Proc. Phys. Soc.* **91** (1967), 475.
86. L. I. Schiff and H. Snyder: *Phys. Rev.* **55** (1939), 59.
87. J. J. Hopfield and D. G. Thomas: *Phys. Rev.* **122** (1961), 35.
88. R. G. Wheeler and J. O. Dimmock: *Phys. Rev.* **125** (1962), 1805.
89. L. I. Schiff and H. Snyder: *Phys. Rev.* **55** (1939), 59.

90. A. G. Zhilich and B. S. Monozon: *Sov. Phys. Solid State* **8** (1967), 2846.
91. A. Baldereschi and F. Bassani: 10th Int. Conf. on Phys. Semicond., 1970, U.S. Atomic Energy Comm., p. 191.
92. L. Fritsche: *Phys. Stat. Sol.* **34** (1969), 195.
93. N. Lee, D. M. Larsen, and B. Lax: *J. Phys. Chem. Solids* **34** (1973), 1059.
94. D. Cabib, E. Fabri, and G. Fiorio: *Solid State Commun.* **9** (1971), 1517.
95. W. S. Boyle and R. E. Howard: *J. Phys. Chem. Solids* **19** (1961), 181.
96. L. Fritsche and F. D. Heidt: *Phys. Stat. Sol.* **35** (1969), 987.
97. J. L. Brebner, J. Halpern, and E. Mooser: *Helv. Phys. Acta* **40** (1967), 385.
98. M. Shinada and S. Sugano: *J. Phys. Soc. Japan* **21** (1966), 1936.
99. O. Akimoto and H. Hasegawa: *J. Phys. Soc. Japan* **22** (1967), 181.
100. M. Shinada and K. Tanaka: *J. Phys. Soc. Japan* **29** (1970), 1258.
101. E. I. Rashba and V. M. Edel'shtein: *Sov. Phys. JETP* **30** (1970), 765.
102. L. M. Roth, B. Lax, and S. Zwerdling: *Phys. Rev.* **114** (1959), 90.
103. E. Burstein, G. S. Picus, R. F. Wallis, and F. Blatt: *Phys. Rev.* **113** (1959), 15.
104. S. O. Sari: *Phys. Rev. Letters* **26** (1971), 1167.
105. A. G. Aronov: *Sov. Phys. Solid State* **5** (1963), 402.
106. Q. H. F. Vrehen: *Phys,. Rev* **145** (1966), 675.
107. K. Ohta: *Phys. Letters (Netherlands)* **26A** (1968), 409.
108. M. Reine, Q. H. F. Vrehen, and B. Lax: *Phys. Rev. Letters* **17** (1966), 582.
109. H. N. Spector: *Phys. Stat. Sol.* **22** (1967), 185.
110. F. Evangelisti and A. Frova: *Phys. Stat. Sol.* **33** (1969), 623.
111. H. N. Spector: *Physica (Netherlands)* **32** (1966), 1551.
112. A. G. Aronov and G. E. Pikus: *Sov. Phys. JETP* **22** (1966), 1300.
113. W. E. Engeler, M. Garfinkel, and J. J. Tiemann: *Phys. Rev.* **155** (1967), 693.
114. C. N. Berglund: *J. Appl. Phys.* **37** (1966), 3019.
115. B. Batz: *Solid State Commun.* **5** (1967), 985.
116. B. O. Seraphin and N. Bottka: *Phys. Rev.* **145** (1966), 628.
117. E. Matatagui, A. G. Thompson, and M. Cardona: *Phys. Rev.* **176** (1968), 950.
118. M. Cardona: *Modulation Spectroscopy*, (Academic Press, New York 1969), p. 119.
119. R. W. G. Wyckoff: *Crystal Structures* **1**, (Inter Science Publ. Inc. New York), Chap. IV.
120. A. Mercier and J. P. Voitchovsky: *Solid State Commun.* **14** (1974), 757.
121. D. J. Lockwood: *Light Scattering Spectra of Solids*, ed. by G. B. Wright, (Springer; Berlin 1969), p. 75.
122. J. P. Mon: *Compt. Rend. Acad. Sci.* **262B** (1966), 493.
123. S. Nakashima, H. Yoshida, and T. Fukumoto: *J. Phys. Soc. Japan* **31** (1971), 1847.
124. C. Carabatos: *Compt. Rend. Acad. Sci.* **272B** (1971), 465.
125. C. G. Trigunayat and G. K. Chada: *Phys. Stat. Sol.* **4** (1971), 9.
126. W. Kleber and P. Fricke: *Z. Phys. Chem. (Leipzig)* **224** (1963), 353.
127. R. S. Mitchell: *Z. Krist.* **108** (1956), 296.
128. M. R. Tubbs: *J. Phys. Chem. Solids* **27** (1966), 1667; and **30** (1969), 2323.
129. P. A. Lee, G. Said, R. Davis, and T. H. Lim: *J. Phys. Chem. Solids* **30** (1969), 2719.
130. D. L. Greenaway and R. Nitsche: *J. Phys. Chem. Solids* **26** (1965), 1445.
131. Ya. O. Dovgii, N. S. Pidzyraylo, M. I. Brilinskii, and S. P. Kudryavets: *Ukr. Fiz. Zh.* **14** (1969), 1804.
132. D. K. Wright and M. R. Tubbs: *Phys. Stat. Sol.* **37**, (1970), 551.
133. S. Brahms: *Phys. Letters (Netherlands)* **19** (1965), 272.
134. M. R. Tubbs: *Phys. Stat. Sol.* **49** (1972), 11.
135. D. M. Adams and M. A. Hooper: *Austr. J. Chem.* **24** (1971), 885.
136. N. Krauzman, M. Krauzman, and H. Poulet: *Compt. Rend. Acad. Sci.* **B273** (1971), 301.
137. J. B. Newkirk: *Acta Metallurgica* **4** (1956), 316.
138. Y. Mikawa, R. J. Jakobsen, and J. W. Brasch: *J. Chem. Phys.* **45** (1966), 4528.
139. Y. Marqueton, F. Abba, E.-A. Decamps, and M.-A. Nusimovici: *Compt. Rend. Acad. Sci.* **B272** (1971), 1014.
140. C. Carabatos: *Opt. Commun.* **1** (1970), 394.
141. K. Kanzaki and I. Imai: *J. Phys. Soc. Japan* **32** (1972), 1003.

142. S. Nikitine: *Prog. Semicond.* **6** (1962), 303.
143. M. Sieskind: *Rev. Opt (France)* **39** (1960), 239.
144. M. Sieskind, J. B. Grun, and S. Nikitine: *J. Phys. Radium (France)* **22** (1961), 777.
145. B. V. Novikov and M. M. Pimonenko: *Fiz. and Tekh. Poluprov (USSR)* **4** (1970), 2077.
146. R. Williams: *Phys. Rev.* **126** (1962), 442.
147. A. Daunois, J. L. Deiss, and S. Nikitine: *Phys. Letters (Netherlands)* **28A** (1968), 274.
148. M. Chester and C. C. Coleman: *J. Phys. Chem. Solids* **32** (1971), 223.
149. R. Kleim, F. Raga, and S. Nikitine: *Proc. Internat. Conf. Luminescence Budapest*, 1968, p. 1496.
150. R. W. G. Wyckoff: *Crystal Structures 1*, John Wiley and Sons Inc., New York 1963, p. 298.
151. G. A. Ozin: *Canad. J. Chem.* **48** (1970), 2931.
152. B. Willemsen: *J. Inorg. Nucl. Chem.* **33** (1971), 3963.
153. K. J. Best: *Z. Phys.* **163** (1961), 309.
154. J. C. Canit: *Phys. Stat. Sol.* **38** (1970), K153.
155. N. L. Kramarenko, V. K. Miloslavskii, and Yu. V. Naboikin: *Ukr. Fiz. Zh.* **15** (1970), 416.
156. G. G. Liidya and V. G. Plekhanov: *Opt. Spectrosc.* **32** (1972), 43.
157. A. F. Malysheva and V. G. Plekhanov: *Opt. Spectrosc.* **34** (1973), 302.
158. R. Hilsch and R. W. Pohl: *Z. Phys.* **48** (1928), 384.
159. H. Fesefeldt: *Z. Phys.* **64** (1930), 741.
160. A. A. Shishlovskii: *Opt. Spectrosc.* **1** (1956), 765.
161. N. M. Gudris and I. N. Shults: *Opt. Spectrosc.* **2** (1957), 246.
162. K. J. De Vries and J. H. van Santen: *Physica* **30** (1964), 2051.
163. R. I. Gindina and A. A. Maaroos: *Opt. Spectrosc.* **26** (1969), 474.
164. N. L. Kramarenko, V. K. Miloslavskii, and Yu. V. Naboikin: *Ukr. Fiz. Zh.* **15** (1970), 416.
165. N. L. Kramarenko and V. K. Miloslavskii: *Phys. Stat. Sol.* **B48** (1971), K177.
166a. G. C. Trigunayat and G. K. Chada: *Phys. Stat. Sol.* **4** (1971), 9.
166b. G. G. Trigunayat: *Phys. Stat. Sol.* **4** (1971), 281.
167. D. L. Greenaway and G. Harbeke: *J. Phys. Soc. Japan* **21** (Supplement), (1966), 151.
168. G. Harbeke and E. Tosatti: *Phys. Rev. Letters* **28** (1972), 1567.
169. For a compilation of experimental data prior to 1972 see [134].
170. Ch. Gahwiller and G. Harbeke: *Phys. Rev.* **185** (1969), 1141.
171. S. Nikitine, J. Schmitt-Burckel, J. Biellmann, and J. Ringeissen: *J. Phys. Chem. Solids* **25** (1964), 951.
172. S. Nikitine and G. Perny: *J. Phys. Radium* **17** (1956), 1017.
173. G. Baldini and S. Franchi: *Phys. Rev. Letters* **26** (1971), 503.
174. J. Hermanson and J. C. Phillips: *Phys. Rev.* **150** (1966), 652.
175. G. Harbeke, F. Bassani, and E. Tosatti; *11th Int. Conf. on Phys. Semiconductors*, Warsaw 1972, p. 163.
176. J. J. Hopfield and D. G. Thomas: *J. Phys. Chem. Solids* **12** (1960), 276.
177. P. I. Perov, L. A. Avdeeva, and M. I. Ellinson: *Sov. Phys. Solid State* **11** (1969), 438.
178. J. Bordas and E. A. Davis: *Solid State Commun.* **12** (1973), 717.
179. D. F. Blossey: *Phys. Rev.* **B3** (1971), 1382.
180. A. J. Grant and A. D. Yoffe: *Phys. Stat. Sol.* (b)**43** (1971), K29.
181. J. B. Anthony and A. D. Brothers: *Phys. Rev.* **B7** (1973), 1539.
182. E. Doni, G. Grosso, and G. Spavierri: *Solid State Commun.* **11** (1972), 493.
183. I. Ch. Schlüter and M. Schluter: *Phys. Rev.* **B9** (1974), 1652.
184. J. Trotter: *Z. Krist.* **121** (1965), 81.
185. J. Trotter and T. Zobel: *Z. Krist.* **123** (1966), 67.
186. S. M. Swingle: private communication, see [187].
187. P. W. Allen and L. E. Sutton: *Acta Cryst.* **3** (1950), 46.
188. M. A. Hooper and D. W. James: *Aust. J. Chem.* **21** (1968), 2379.
189. T. R. Manley and D. A. Williams: *Spectrochim. Acta* **21** (1965), 1773.
190. W. Kiefer: *Z. Naturforsch.* **25A** (1970), 1101.
191. M. A. Hooper and D. W. James: *Spectrochim. Acta* **25A** (1969), 569.
192. V. I. Vashshenko, V. N. Kolosyuk, and G. A. Vashchenko: *Sov. Phys. – Solid State* **14** (1973), 3068.
193. M. R. Tubbs: *J. Phys. Chem. Solids* **29** (1968), 1191.

194. V. I. Vashchenko, V. B. Timofeev, and I. N. Antipov: *Optics Spectrosc.* **22** (1967), 440.
195. B. L. Evans: *Proc. Roy. Soc.* **A289** (1966), 275.
196. B. L. Evans: *Proc. Roy. Soc.* **A276** (1963), 136.
197. V. I. Vashchenko and V. B. Timofeev: *Sov. Phys.–Solid State* **9** (1967), 1242.
198. E. F. Gross, V. I. Perel, and R. I. Shekhmamet'ev: *JETP Letters* **13** (1971), 229.
199. E. F. Gross, D. L. Federov, and R. I. Shekhmamet'ev: *Sov. Phys.–Solid State* **14** (1973), 2767.
200. V. B. Timofeev and V. I. Vashchenko: *Optics Spectrosc.* **24** (1968), 396.
201. A. M. Vainrub, A. V. Il'inskii, and B. V. Novikov: *Sov. Phys.–Solid State* **15** (1973), 343.
202. Y. Petrov, P. Y. Yu, and Y. R. Shen: *Phys. Stat. Sol.* (b)**61** (1974), 419.
203. V. N. Kolosyuk and V. I. Vaschenko: *Opt. Spectrosc.* **34** (1973), 236.
204. R. Pappalardo: *J. Chem. Phys.* **31** (1959), 1050.
205. R. Pappalardo: *J. Chem. Phys.* **33** (1960), 613.
206. J. W. Stout: *J. Chem. Phys.* **33** (1960), 303.
207. S. Sato *et al.*: *J. Phys. Soc. Japan* **30** (1971), 459.
208. T. Ishii, Y. Sakisaka, T. Matsukawa, S. Sato, and T. Sagawa: *Solid State Commun.* **13** (1973), 281.
209. A. Roth and S. Robin: *Solid State Commun.* **14** (1974), 673.
210. A. Trutia and M. Musa: *Phys. Stat. Sol.* **8** (1965), 663.
211. S. Koide and M. H. L. Pryce: *Phil. Mag.* **3** (1958), 607.
212. Y. Tanabe and S. Sugano: *J. Phys. Soc. Japan* **9** (1954), 753.
213. L. L. Lohr: *J. Chem. Phys.* **45** (1966), 3611.
214. J. C. Zahner and H. G. Drickamer: *J. Chem. Phys.* **35** (1961), 1483.
215. J. Ferguson, D. L. Wood, and K. Knox: *J. Chem. Phys.* **39** (1963), 881.
216. Z. S. Basinski, D. B. Dove, and E. Mooser: *Helv. Phys. Acta* **34** (1961), 373.
217. W. B. Pearson: *Acta Cryst.* **17** (1964), 1.
218. L. Pauling: *The Nature of the Chemical Bond*, (Cornell University Press, Ithaca, New York 1960).
219. T. J. Wieting and J. L. Verble: *Phys. Rev.* **B5** (1972), 1473.
220. J. P. van der Ziel, A. E. Meixner, and H. M. Kasper: *Solid State Commun.* **12** (1973), 1213.
221. P. C. Leung, G. Anderson, W. G. Spitzer, and C. A. Mead: *J. Phys. Chem. Solids* **27** (1966), 849.
222. T. J. Wieting: *Solid State Commun.* **12** (1973), 937.
223. G. A. Akhundov and T. G. Kerimova: *Optics Spectrosc.* **22** (1967), 355.
224. N. Kuroda, Y. Nishina, and T. Fukuroi: *J. Phys. Soc. Japan* **24** (1968), 214.
225. H. Yoshida, S. Nakashima, and A. Mitsuishi: *Phys. Stat. Sol.* (b)**59** (1973), 655.
226. M. Hayek, O. Brafman, and R. M. A. Lieth: *Phys. Rev.* **B8** (1973), 2772.
227. J. C. Irwin, R. M. Hoff, B. P. Clayman, and R. A. Bromley: *Solid State Commun.* **13** (1973), 1531.
228. E. F. Gross, B. V. Novikov, B. S. Razbirin, and L. G. Suslina: *Optics Spectrosc.* **6** (1959), 364.
229. J. L. Brebner and G. Fischer: *Canad. J. Phys.* **41** (1963), 561.
230. M. A. Nizametdinova: *Phys. Stat. Sol.* **19** (1967), K111.
231. J. L. Brebner: *J. Phys. Chem. Solids* **25** (1964), 1427; also *Helv. Phys. Acta* **37** (1964), 589.
232. F. I. Ismailov, E. S. Guseinova, and G. A. Akhundov: *Sov. Phys. Solid State* **5** (1964), 2656.
233. G. Fischer: *Helv. Phys. Acta* **36** (1963), 317.
234. H. Kamimura and K. Nakao: *J. Phys. Soc. Japan* **24** (1968), 1313.
235. E. Aulich, J. L. Brebner, and E. Mooser: *Phys. Stat. Sol.* **31** (1969), 129.
236. Y. Nishina, N. Kuroda, and T. Fukuroi: *Proc. Int. Conf. on Phys. Semiconductors Moscow*, (Nauka, Leningrad 1968), p. 1024.
237. H. Kamimura, K. Nakao, and Y. Nishina: *Phys. Rev. Letters* **22** (1969), 1379.
238. K. Nakao, H. Kamimura, and Y. Nishina: *Nuovo Cimento* **B63** (1969), 45.
239. J. M. Besson, K. P. Jain, and A. Kuhn: *Phys. Rev. Letters* **32** (1974), 936.
240. H. I. Ralph: *Phys. Letters (Netherlands)* **27A** (1968), 431.
241. M. Grandolfo, F. Somma, and P. Vecchia: *Phys. Rev.* **B5,** (1972), 428.
242. F. Bassani, D. L. Greenaway, and G. Fischer: *Proc. VII Int. Conf. on Phys. Semicond. Paris 1964*, (Academic Press, New York 1965).

243. S. Kohn, Y. Petroff, and Y. R. Shen: *Surface Science* **37** (1973), 205.
244. J. Halpern: *J. Phys. Soc. Japan* **21,** Suppl. (1966), 180.
245. J. L. Brebner and E. Mooser: *Phys. Letters* **24A** (1967), 274.
246. E. Mooser and M. Schlüter: *Nuovo Cimento* **18B,** Ser. 2 (1973), 164.
247. M. Schlüter: *Nuovo Cimento* **13B,** Ser. 2 (1973), 313.
248. J. L. Brebner and E. Mooser: *Helv. Phys. Acta* **38** (1965), 656.
249. M. I. Karaman and V. P. Mushinskii: *Fiz. Tekh. Poluprov.* **4** (1970), 1143 and 424.
250. Yu. D. Dumarevskii *et al.*: *Phys. Stat. Sol.* **39** (1970), K71.
251. V. A. Gadzhiev *et al.*: *Fiz. Tverd. Tela* **12** (1970), 1350.
252. Y. Suzuki *et al.*: *J. Phys. Chem. Solids* **31** (1970), 2217.
253. A. Balzarotti, M. Piacentini, E. Burattini, and P. Picozzi: *J. Phys.* **C4** (1971), L273.
254. A. Balzarotti, M. Grandolfo, F. Somma, and P. Vecchia: *Phys. Stat. Sol.* (b)**44** (1971), 713.
255. M. Grandolfo, E. Gratton, F. Anfosso Somma, and P. Vecchia: *Phys. Stat. Sol.* (b)**48** (1971), 729.
256. S. Antoci and L. Mihich: *Solid State Commun.* **12** (1973), 649.
257. J. L. Brebner: *J. Phys. Chem. Solids* **25** (1964), 1427.
258. A. Bourdon and F. Khelladi: *Solid State Commun.* **9** (1971), 1715.
259. M. Shinada and S. Sugano: *J. Phys. Soc. Japan* **20** (1965), 1274.
260. Y. Nishina, S. Kurita, and S. Sugano: *J. Phys. Soc. Japan* **21** (1966), 1609.
261. J. L. Brebner, J. Halpern, and E. Mooser: *Helv. Phys. Acta* **40** (1967), 385.
262. J. L. Brebner: *Canad. J. Phys.* **51** (1973), 497.
263. D. Bianchi, U. Emiliani, P. Podini, and C. Paorici: *Phys. Stat. Sol.* (b)**60** (1973), 511.
264. P. C. Leung, G. Anderman, W. G. Spitzer, and C. A. Mead: *J. Phys. Chem. Solids* **27** (1966), 849.
265. J. A. Deverin: *Helv. Phys. Acta* **42** (1969), 397.
266. R. Mamy, L. Martin, G. Leveque, and C. Raisin: *Phys. Stat. Sol.* (b)**62** (1974), 201.
267. V. V. Sobolev and V. I. Donetskich: *Phys. Stat. Sol.* (b)**45** (1971), K15.
268. A. Divrechi, C. Ance, and J. Robin: *Compt. Rend. Acad. Sci.* **B274** (1972), 815.
269. T. Lane, C. J. Vesely, and D. W. Langer: *Phys. Rev.* **B6** (1973), 3770.
270. W. Gudat *et al.*: *Phys. Stat. Sol.* (b)**52** (1972), 505.
271. M. Cardona, C. M. Penchina, N. J. Shevchik, and J. Tejada: *Solid State Commun.* **11** (1972), 1655.
272. R. Vilanove: Thesis, Paris 1971, CNRS No. A05885.
273. N. A. Gasanova, G. A. Akhundov, and M. A. Nizametdinova: *Phys. Stat. Sol.* **17** (1966), K115.
274. V. K. Subashiev, Le Hak Bin, and L. S. Chertkova: *Solid State Commun.* **9** (1971), 369.
275. A. Balzarotti and M. Piacentini: *Solid State Commun.* **10** (1972), 421.
276. V. I. Sokolov, D. B. Kushev, and V. K. Subashiev: *Phys. Stat. Sol.* (b)**50** (1972), K125.
277. Y. Sasaki, C. Hamaguchi, and J. Nakai: *J. Phys.* **C5** (1972), L95.
278. J. L. Brebner and G. Fischer: *1962 Semiconductor Conference Exeter*, (Inst. of Phys., 1962), p. 760.
279. N. A. Gasanova and G. A. Akhundov: *Opt. Spectrosc.* **18** (1965), 413.
280. S. M. Ryvkin and R. I. Khansevarov: *Zh. Tekh. Fiz.* **26** (1956), 2781.
281. P. Fielding, G. Fischer, and E. Mooser: *J. Phys. Chem. Solids* **8** (1959), 434.
282. J. L. Brebner, G. Fischer, and E. Mooser: *J. Phys. Chem. Solids* **23** (1962), 1417.
283. C. Tatsuyama, Y. Watanabe, C. Hamaguchi, and J. Nakai: *J. Phys. Soc. Japan* **29** (1970), 150.
284. D. Gili-Tos, M. Grandolfo, and P. Vechia: *Phys. Rev.* **B7** (1973), 2565.
285. E. Burattini, M. Grandolfo, G. Mariutti, and C. Ranghiasci: *Surface Science* **37** (1973), 198.
286. M. L. Belle and N. A. Gasanova: *Opt. Spectrosc.* **18** (1965), 412.
287. V. Grasso, G. Mondio, and G. Saitta: *Phys. Letters* **46A** (1973), 95.
288. V. Grasso, G. Mondio, M. A. Pirrone, and G. Saitta: *J. Phys.* **C8** (1975), 80.
289. F. Consadori and J. L. Brebner: *Solid Stat. Commun.* **12** (1973), 179.
290. H. Krebs: *Fundamentals of Inorganic Crystal Chemistry*, (McGraw-Hill, London 1968).
291. H. Schäfer: *Chemical Transport Reactions*, (Academic Press, New York 1964).
292. D. L. Greenaway and R. Nitsche: *J. Phys. Chem. Solids* **26** (1965), 1445.
293. V. V. Sobolev and V. I. Donetskich: *Phys. Stat. Sol.* **42** (1970), K53.
294. G. Domingo, R. S. Itoga, and C. R. Kannewurf: *Phys. Rev.* **143** (1966), 536.

295. P. A. Lee, G. Said, R. Davis, and T. H. Lim: *J. Phys. Chem. Solids* **30** (1969), 2719.
296. P. A. Lee and G. Said: *J. Phys. D, Ser 2*, **1** (1968), 837.
297. B. L. Evans and R. A. Hazelwood: *J. Phys.* **D2** (1969), 1507.
298. F. Aymerich, F. Meloni, and G. Mula: *Solid State Commun.* **12** (1973), 139.
299. B. L. Evans and J. D. Bradley: (unpublished).
300. S. Asanabe: *J. Phys. Soc. Japan* **16** (1961), 1789.
301. G. Busch, C. Frolich, and F. Hulliger: *Helv. Phys. Acta* **34** (1961), 359.
302. G. Mula and F. Aymerich: *Phys. Stat. Sol.* (b)**51** (1972), K35.
303. M. Y. Au-Yang and M. L. Cohen: *Phys. Rev.* **178** (1969), 1279.
304. C. Y. Fong and M. L. Cohen: *Phys. Rev.* **B5** (1972), 3095.
305. A. A. Vaipolin: *Sov. Phys. Cryst.* **10** (1966), 509.
306. M. J. Buerger: *Amer. Min.* **27** (1942), 301.
307. R. Zallen, M. L. Slade, and A. T. Ward: *Phys. Rev.* **B3** (1971), 4257.
308. B. L. Evans and P. A. Young: *Proc. Roy. Soc.* **A297** (1967), 1449.
309. M. T. Kostyshin and P. F. Romanenko: *Ukr. Fiz. Zh.* (*USSR*) **8** (1963), 102.
310. M. P. Lisitsa, L. I. Berezhinsky, M. Ya. Valakh, and A. M. Yaremko: *Phys. Letters* **42A** (1972), 51.
311. M. T. Kostyshin and P. F. Romanenko: *Ukr. Fiz. Zh.* (*USSR*). **9** (1964), 166.
312. I. S. Gorban and R. A. Dashkovskaya: *Sov. Phys.* (*Solid State*) **6** (1965), 1895.
313. R. F. Shaw, W. Y. Liang, and A. D. Yoffe: *J. Non-Cryst. Solids* **4** (1970), 29.
314. R. Zallen, R. E. Drews, R. L. Emerald, and M. L. Slade: *Phys. Rev. Letters* **26** (1971), 1564.
315. R. E. Drews, R. L. Emerald, M. L. Slade, and R. Zallen: *Solid State Commun.* **10** (1972), 293.
316. J. Perrin, J. Cazaux, and P. Soukiassian: *Phys. Stat. Sol.* (b)**62** (1974), 343.
317. V. Riede: *Ann. Phys.* (*Germany*) **25** (1970), 415.
318. B. H. Billings and M. Hyman: *J. Opt. Soc. Amer.* **37** (1947), 119.
319. A. S. Valeev and M. A. Gisin: *Opt. Spectrosc.* **19** (1965), 62.
320. J. Kucirek: *Czech. J. Phys.* **B18** (1968), 795.
321. A. Yul, Shileika Lietuvos, and T. S. R. Mokslu: *Akad. Darbai. Ser B1*, **21** (1960), 107.
322. M. T. Kostyskin: *Opt. Spektrosc.* **5** (1958), 71.
323. A. F. Skubenko and S. V. Lapshii: *Sov. Phys.– Solid State* **4** (1962), 327.
324. A. I. Audzionis and A. S. Karpus: *Sov. Phys.– Solid State* **11** (1969), 859.
325. K. A. Verkhovskaya, I. P. Grigas, and V. M. Fridkin: *Sov. Phys.– Solid State* **10** (1969), 1583.
326. A. G. Khasabov and I. Ya. Nikiforov: *Sov. Phys. Crystallog.* **16** (1971), 28.
327. W. Procarione and C. Wood: *Phys. Stat. Sol.* **42** (1970), 871.
328. A. F. Skubenko and S. V. Laptyi: *Ukranian. Fiz. Zh.* (*USSR*) **9** (1964), 744.
329. B. Van Pelt and C. Wood: referred to in [331].
330. V. V. Sobolev, S. D. Shutov, and S. N. Shestatskii: *Moldavian Acad. Sci.* (1969). 183.
331. J. C. Shaffer, B. Van Pelt, C. Wood, J. Freeouf, K. Murase, and J. W. Osmun: *Phys. Stat. Sol.* (b)**54** (1972), 511.
332. S. D. Shutov, V. V. Sobolev, Y. V. Popov, and S. N. Shestatskii: *Phys. Stat. Sol.* **31** (1969), K23.
333. E. Mooser and W. B. Pearson: *J. Phys. Chem. Solids* **7** (1958), 65.
334. N. W. Tideswell, F. H. Kruse, and J. C. McCullough: *Acta Cryst.* **10** (1957), 99.
335. H. Kohler and C. R. Becker: *Phys. Stat. Sol.* (b)**61** (1974), 533.
336. K. H. Unkelbach, Ch. Becker, H. Kohler, and A. V. Middendorff: *Phys. Stat. Sol.* (b)**60** (1973), K41.
337. K. Taniguchi, A. Moritani, C. Hamaguchi, and J. Nakai: *Surface Science* **37** (1973), 212.
338. A. Balzarotti, E. Burattini, and P. Picozzi: *Phys. Rev.* **B3** (1971), 1159.
339. U. Giorgianni, V. Grasso, and G. Saitta: *Lett. Nuovo Cimento* **5** (1972), 951.
340. V. V. Sobolev, N. N. Syrbu, V. K. Nikitina, and Yu. K. Lubanova: *Phys. Stat. Sol.* **42** (1970), K85.
341. D. L. Greenaway and G. Harbeke: *J. Phys. Chem. Solids* **26** (1965), 1585.
342. V. V. Sobolev, S. D. Shutov, Yu. V. Porov, and S. N. Shestatskii: *Phys. Stat. Sol.* **30** (1968), 349.
343. S. Katsuki: *J. Phys. Soc. Japan* **26** (1969), 58.
344. F. Borhese and E. Donato: *Nuovo Cimento* **53B** (1968), 283.
345. V. V. Sobolev: *Phys. Stat. Sol.* (b)**49** (1972), K29.

346. R. Sehr and L. R. Testardi: *J. Phys. Chem. Solids* **23** (1962), 1219.
347. I. Termoto and S. Takayanagi: *J. Phys. Chem. Solids* **19** (1961), 124.
348. K. Schubert, K. Anderko, M. Kluge, H. Buskow, M. Ilschner, E. Dorre, and P. Esslinger: *Naturwiss.* **40** (1953), 269.
349. H. Gobrecht, S. Seeck, and T. Klose: *Z. Phys.* **190** (1966), 427.
350. I. G. Austin: *Proc. Phys. Soc.* **72** (1968), 545.
351. I. G. Austin and A. Sheard: *J. Electron. Contr.* **3** (1957), 236.
352. J. R. Drabble and C. H. L. Goodman: *J. Phys. Chem. Solids* **5** (1958), 142.
353. R. Hultgren: *Phys. Rev.* **40** (1932), 891.
354. F. Jellinek: *Ark. Kemi* **20**/36 (1962), 447.
355. H. Haraldsen: *Angew. Chem. Internat. Edit. English* **5** (1), (1966), 58.
356. B. B. Zvyagin and S. V. Soboleva: *Sov. Phys. Cryst.* **12** (1967), 46.
357. B. E. Brown: *Acta Crystallog.* **20** (1966), 264.
358. N. W. Alcock and A. Kjekshus: *Acta Chem. Scand.* **19** (1965), 79.
359. J. L. Verble and T. J. Wieting: *Phys. Rev. Letters* **25** (1970), 362.
360. G. Lucovsky, R. M. White, J. A. Benda, and J. F. Revelli: *Phys. Rev.* **B7,** (1973), 3859.
361. P. J. Lockward: *Light Scattering in Solids*, ed. by G. B. Wright, (Springer, New York 1969), p. 75.
362. T. J. Wieting and J. L. Verble: *Phys. Rev.* **B3** (1971), 4286.
363. J. E. Smith, M. I. Nathan, M. W. Shafer, and J. B. Torrance: *11 Int. Conf. on Phys. of Semiconductors*, Warsaw 1972, p. 1306.
364. O. P. Agnihotri and H. K. Sehgal: *Phil. Mag.* **26** (1972), 753.
365. A. K. Garg, H. K. Sehgal, and O. P. Agnihotri: *Solid State Commun.* **12** (1973), 1261.
366. A. Gleizes and Y. Jeannin: *J. Solid State Chem.* **5** (1972), 4.
367. J. M. Chen and C. S. Wang: *Solid State Commun.* **14** (1974), 857.
368. O. P. Agnihotri, H. K. Sehgal, and A. K. Garg: *Solid State Commun.* **12** (1973), 135.
369. J. L. Verble and T. J. Wieting: *Solid State Commun.* **11** (1972), 941.
370. L. V. Azaroff and J. J. Brophy: *Electronic Processes in Materials*, (McGraw-Hill, New York 1963).
371. L. Pauling: *The Nature of the Chemical Bond*, (Cornell University Press, 1960).
372. R. M. White and G. Lucovsky: *Solid State Commun.* **11** (1972), 1369.
373. J. A. Wilson and A. D. Yoffe: *Adv. Phys.* **18** (1969), 193.
374. L. E. Conroy and K. C. Park: *Inorg. Chem.* **7** (1968), 459.
375. D. L. Greenaway and R. Nitsche: *J. Phys. Chem. Solids* **26** (1965), 1445.
376. P. A. Lee, G. Said, R. Davis, and T. H. Lim: *J. Phys. Chem. Solids* **30** (1969), 2719.
377. A. R. Beal, J. C. Knights, and W. Y. Liang: *J. Phys.* **C5** (1972), 3531.
378. A. J. Grant, J. A. Wilson, and A. D. Yoffe: *Phil. Mag.* **25** (1972), 625.
379. A. R. Beal and W. Y. Liang: *Phil. Mag.* **27** (1973), 1397.
380. R. Vilanove: *C. R. Hebd Sean, Acad. Sci.* **B271** (1970), 1101.
381. W. Y. Liang and S. L. Cundy: *Phil. Mag.* **19** (1969), 1031.
382. A. Couget, L. Martin, and F. Pradal: *C. R. Hebd. Sean. Acad. Sci.* **B272** (1971), 626.
383. R. B. Murray, R. A. Bromley, and A. D. Yoffe: *J. Phys.* **C5** (1972), 746.
384. R. B. Murray and A. D. Yoffe: *J. Phys.* **C5** (1972), 3038.
385. R. F. Frindt: *Phys. Rev. Letters* **28** (1972), 299.
386. B. P. Clayman and R. F. Frindt: *Solid State Commun.* **9** (1971), 1881.
387. L. Martin, R. Mamy, A. Couget, and C. Raisin: *Phys. Stat. Sol.* (b)**58** (1973), 623.
388. C. Leveque, S. Robin-Kandare, L. Martin, and F. Pradal: *Phys. Stat. Sol.* (b)**58** (1973), K65.
389. W. Y. Liang: *J. Phys.* **C4** (1971), L378.
390. F. Consadori and R. F. Frindt: *Solid State Commun.* **9** (1971), 2151.
391. W. Y. Liang: *J. Phys.* **C6** (1973), 551.
392. A. Couget, L. Martin, F. Pradal, and R. Nitsche: *Phys. Letters* **41A** (1972), 261.
393. L. F. Matheiss: *Phys. Rev. Letters* **30** (1973), 784.
394. D. J. Bradley, Y. Katayama, and B. L. Evans: *Solid State Commun.* **11** (1972), 1695.
395. P. M. Williams and F. R. Shepherd: *J. Phys.* **C6** (1973), L36.
396. R. H. Williams, J. M. Thomas, M. Barber, and N. Alford: *Chem. Phys. Letters* **17** (1972), 142.

397. R. S. Title and M. W. Shafer: *Phys. Rev. Letters* **28** (1972), 808.
398. E. A. Antonova, V. G. Vorob'ev, G. A. Kalyuzhnaya, and V. V. Sobolev: *Fiz. Tekh. Poluprov* **3** (1969), 922.
399. V. V. Sobolev, V. I. Donetskii, G. A. Kalyuzhnaya, and E. A. Antonova: *Fiz. Tekh. Poluprov.* **5** (1971), 959.
400. R. F. Frindt and A. D. Yoffe: *Proc. Roy. Soc.* **A273** (1963), 69.
401. B. L. Evans and P. A. Young: *Proc. Roy. Soc.* **A284** (1965), 402.
402. V. V. Sobolev: *Optics Spectrosc.* **18** (1965), 187.
403. R. Bailly: *Amer. Min.* **33** (1948), 519.
404. For a review see: A. D. Yoffe: *Festkorper Probleme* **13** (1973), 1.
405. R. A. Neville and B. L. Evans: *Phys. Stat. Sol.* (b)**73** (1976) 597.
406. B. L. Evans and R. A. Hazelwood: *Phys. Stat. Sol.* **A4** (1971), 181.
407. B. Davey and B. L. Evans: *Phys. Stat. Sol.* **A13** (1972), 483.
408. A. A. Al-Hilli and B. L. Evans: *J. Cryst. Growth* **15** (1972), 93.
409. B. L. Evans and P. A. Young: *Phys. Stat. Sol.* **25** (1968), 417.
410. B. L. Evans and P. A. Young: *Proc. Roy. Soc.* **A298** (1967), 74.
411. P. G. Harper and J. A. Hilder: *Phys. Stat. Sol.* **26** (1968), 69.
412. G. Jones and J. L. Brebner: *J. Phys.* **C4** (1971), 723.
413. R. F. Frindt: *Phys. Rev.* **140** (1965), A536.
414. A. Clark and R. H. Williams: *J. Phys.* **D1** (1968), 1222.
415. G. A. N. Connell, J. A. Wilson, and A. D. Yoffe: *J. Phys. Chem. Solids* **30** (1969), 287.
416. G. Weiser: *Surface Science* **37** (1973), 175.
417. J. Bordas and E. A. Davis: *Phys. Stat. Sol.* (b)**60** (1973), 505.
418. B. L. Evans and P. A. Young: *Proc. Phys. Soc.* **91** (1967), 475.
419. V. V. Sobolev, V. I. Donetskikh, A. A. Opalovskii, V. E. Federov, E. U. Lobkov, and A. P. Mazhara: *Sov. Phys. Semicond.* **5** (1971), 909.
420. F. R. Gamble: *J. Solid State Chem.* **9** (1974), 358.
421. K. Zeppenfeld: *Optics Commun.* **1** (1970), 377.
422. J. A. Bearden and A. F. Burr: *Rev. Mod. Phys.* **39** (1967), 125.
423. H. P. Hughes and W. Y. Liang: *J. Phys.* **C7** (1974), 1023.
424. V. Grasso, G. Mondio, and G. Saitta: *J. Phys.* **C5** (1972), 1101.
425. R. A. Bromley, R. B. Murray, and A. D. Yoffe: *J. Phys.* **C5** (1972), 759.
426. P. G. Harper and D. R. Edmondson: *Phys. Stat. Sol.* **44** (1971), 59.
427. R. Huisman, R. de Jonge, C. Haas, and F. Jellinek: *J. Solid State Chem.* **3** (1971), 56.
428. R. V. Kasowski: *Phys. Rev. Letters* **30** (1973), 1175.
429. J. C. McMenamin and W. E. Spicer: *Phys. Rev. Letters* **29** (1972), 1501.
430. R. S. Title and M. W. Shafer: *Phys. Rev. Letters* **28** (1972), 808; also *Phys. Rev.* **B8** (1973), 615.
431. S. Antoci, P. Camagni, A. Manara, and A. Stella: *J. Phys. Chem. Solids* **33** (1972), 1177.
432. F. Consadori and R. F. Frindt: *Phys. Rev.* **B2** (1970), 4893.
433. G. Jones: *Phys. Rev.* **B4** (1971), 2069.
434. G. Campagnoli, G. Giuliani, A. Gustinetti, and A. Stella: *Solid State Commun.* **11** (1972), 945.
435. G. Guizzetti, L. Nosenzo, E. Reguzzoni, and G. Samoggia: *Phys. Rev. Letters* **31** (1973), 154.
436. B. Feuerbacher and B. Fitton: *Phys. Rev. Letters* **26** (1971), 840.
437. J. Zupan: *Phys. Rev.* **B6** (1972), 2477.
438. G. S. Painter and D. E. Ellis: *Phys. Rev.* **B1** (1970), 4747.
439. W. van Haeringen and H. G. Junginger: *Solid State Commun.* **7** (1969), 1723.
440. E. Doni and G. Pastori Parravicini: *Nuovo Cimento* **64B** (1969), 117.
441. H. Lipson and A. R. Stokes: *Proc. Roy. Soc.* **A181** (1942), 101.
442. R. R. Haering: *Canad. J. Phys.* **36** (1958), 352.
443. J. W. McClure: *Carbon* **7** (1969), 425.
444. E. A. Taft and H. R. Philipp: *Phys. Rev.* **138** (1965), A197.
445. Y. Sato: *J. Phys. Soc. Japan* **24** (1968), 489.
446. D. L. Greenaway, G. Harbeke, F. Bassani, and E. Tosatti: *Phys. Rev.* **178** (1969), 1340.
447. S. Ergun, J. B. Yasinsky, and J. R. Townsend: *Carbon* **5** (1967), 403.
448. E. Tosatti and F. Bassani: *Nuovo Cimento* **65B** (1970), 161.

449. K. Zeppenfeld: *Z. Phys.* **211** (1968), 391.
450. T. Uda: *J. Phys. Soc. Japan* **28** (1970), 946.
451. See F. Bassani and G. Pastori Parravicini: *Nuovo Cimento* **B50** (1967), 95; which labels the B.Z differently to Figure 68.
452. J. Cazaux: *Solid State Commun.* **8** (1970), 545.
453. G. Harbeke and E. Tosatti: *RCA Review* **36** (1975), 40.
454. A. Roth: Thesis L'Université de Rennes (1974).
455. H. Kamimura and K. Nakao: (1976),–private communication.
456. Y. Hamakawa, P. Handler and F. A. Germano: *Phys. Rev.* **167** (1968), 709.

Acknowledgements

The author is grateful to:-

The American Institute of Physics for permission to reproduce the data given in Figs 11 [456], 14 [72], 24 [168], 25 [170], 30 [237], 32 [241], 35 [241], 45 [338], 49 [360], 50 [360, 362], 55 [393], 62 [428], 67 [432], 68 [438], 69 [446].

The Institute of Physics for permission to reproduce the data given in Figs. 35 [277], 36 [288], 37b [297], 51 [377], 53 [383], 54b [391], 61 [423], 63 [391], 64 [428].

Societa Italiana di Fisica for permission to reproduce the data given in Figs. 12 [63], 33a [246], 45 [339].

Academic-Verlag for permission to reproduce the data given in Figs. 1 [6], 2 [6], 5 [6], 6 [6], 7 [20], 8 [20], 9 [33], 18a [132], 30 [255], 33c [254], 35 [266], 37a [293], 38 [302], 41 [316], 42 [331, 332], 43 [342], 44 [340], 54a [387, 388], 56 [405, 406, 407], 57 [405, 409], 58 [417], 59 [405], 60 [387, 388], 65 [407].

The Japanese Institute of Physics for permission to reproduce the data given in Figs. 10 [34], 19 [141], 20 [141], 21 [141], 24 [167], 31b [234], 46a [343], 69 [445].

The Royal Society for permission to reproduce the data given in Figs. 27 [195], 39 [308].

Taylor and Francis Ltd., for permission to reproduce the data given in Figure 54c [379].

North Holland Publishing Co. for permission to reproduce the data given in Figs. 32 [243], 35 [243], 45 [337], 58 [416], 16b [240], 18b [133], 36 [287], 54b [392].

National Research Council of Canada for permission to reproduce the data given in Figs. 30 [229], 33b [262], 34 [262].

Pergamon Press Ltd., for permission to reproduce the data given in Figs. 16a [82], 18b [292], 27 [193], 28 [193], 37a [292], 43 [341], 52 [292], 28 [208], 29 [208, 209], 35 [275], 36 [289], 37c [298], 40 [315], 66 [394].

Academic Press Inc. for permission to reproduce the data given in Figure 17 [118]. The publishers of *J. Solid State Chem.* for permission to reproduce the data given in Figure 47c [420].

SOME ASPECTS OF MODULATION SPECTROSCOPY IN LAYER MATERIALS

J. BORDAS

Cavendish Laboratory, University of Cambridge, England

1. INTRODUCTION	145
2. OPTICAL ABSORPTION BY SOLIDS	147
3. ON THE THEORY OF ELECTROMODULATION	154
4. EXPERIMENTAL TECHNIQUES OF ELECTROMODULATION	163
5. A GENERAL MODEL FOR ELECTROMODULATION SIGNALS	166
6. ELECTROMODULATION EXPERIMENTS ON LAYER MATERIALS	171
7. ELECTROMODULATION SPECTROSCOPY OF EXCITONS AT THE FUNDAMENTAL ABSORPTION EDGE IN THE HIGH FIELD REGIME	172
7.1. The exciton at the fundamental absorption edge of PbI_2	172
7.2. A model for electric field effects on biaxial excitons in the high field regime	179
8. THE EXCITONS AT THE FUNDAMENTAL ABSORPTION EDGE OF GaS, GaSe, AND GaTe	181
9. ELECTROMODULATION SPECTROSCOPY OF EXCITONS AT THE FUNDAMENTAL ABSORPTION EDGE IN THE LOW FIELD REGIME	187
9.1. Electromodulation experiments on the A exciton at the fundamental absorption edge of $2H-MoS_2$	189
10. ELECTROABSORPTION ON THE A SERIES OF $3R-MoS_2$, $3R-WS_2$, $2H-WSe_2$, $2H-MoS_2$	199
11. DEVELOPMENT OF THE EA LINE SHAPES FROM THE LOW FIELD TO THE HIGH FIELD REGIME	205
12. SUMMARY OF ELECTROMODULATION EXPERIMENTAL RESULTS	206
13. ELECTROMODULATION RESULTS FOR PHOTON ENERGIES ABOVE THE FUNDAMENTAL ABSORPTION EDGE	207
13.1. Electromodulation experiments on GaSe above the fundamental absorption edge	207
13.2. Electromodulation experiments on some layer materials for photon energies above the fundamental absorption edge	208
14. EM EXPERIMENTS ON VERY ANISOTROPIC LAYER MATERIALS - As_2Se_3	214
15. THERMOMODULATION TECHNIQUES	220
15.1. Thermomodulation experiments on some layer materials (GaSe, GaTe, Graphite, GaS	222
16. FINAL REMARKS	226
REFERENCES	227

1. Introduction

The optical spectra of solids very rarely show extremely sharp features as obtained in spectroscopy of nearly isolated atoms. This is because of the high particle density of solids, which leads to overlap of the atomic orbitals and to the formation of energy bands, making their optical spectra rather broad, with any structure usually superimposed on a large structureless background. The lack of sharp structure is more noticeable, of course, in the case of materials with metallic and covalent bonding than for those with ionic bonding because of the smaller atomic orbital overlap in the latter. Experimental information about transitions in solids is important in order to understand how the many body interactions lead to the formation of their electronic band structure.

P. A. Lee (ed.), Optical and Electrical Properties, 145–229. All Rights Reserved.
Copyright © 1976 by D. Reidel Publishing Company, Dordrecht-Holland.

Modulation techniques have become a powerful tool to serve this purpose. The basis of modulation spectroscopy lies in looking for the induced changes in the transmission (T) or reflection (R) of a solid when some periodic external perturbation is applied to it. Depending on what type of perturbation is applied, different terms are used to describe the technique and one has, for example, thermomodulation, piezomodulation, photomodulation, wavelength modulation and electromodulation when the period perturbation is temperature, pressure, light, wavelength or an electric field respectively [1].

From the experimentalist point of view one of the advantages of modulation spectroscopy is that one measures a derivative-like signal, consequently obtaining a considerable enhancement in the structure of the optical spectra. In addition, one has the advantage of using phase sensitive detecting techniques with which one can improve the signal-to-noise ratio and the production of automatically normalized derivative-like spectra. Nevertheless, modulation techniques can offer a lot more than merely a technique to detect hidden structure and, ever since Seraphin and Hess [2] performed the first electro-reflection measurements, a number of new uses for modulation techniques has been found, some of them of particular importance when studying layer materials. An obvious application is to make use of the availability of a perturbation which one can apply along chosen directions of the crystal (pressure or electric fields) and which will reflect the anisotropic nature of the lattice in the modulation spectra.

Most layer materials can be cleaved to the required thicknesses to perform modulated transmission experiments which have certain advantages over reflection experiments: (a) the quality of the surfaces is not so critical; (b) in electromodulation at least the relative change in transmission ($\Delta T/T$) is at least usually one order of magnitude bigger than the relative change in reflection ($\Delta R/R$) and (c), which is of particular importance, the response obtained is frequently more directly related to the imaginary part of the dielectric constant which is the quantity usually calculated by theoreticians.

Of all modulation techniques, electromodulation (EM) has become the most commonly used, possibly because of its relative simplicity. The popularity of EM is reflected in the modulation work done on layer materials, for which the main bulk of published data is concerned with it. Thermomodulation (TM) comes out as a second best, while other techniques have been used only very sparsely to date.

However, despite the fact that EM is a widely used technique and the theoretical groundwork was laid sixteen years ago by Franz and Keldysh [3, 4], who independently performed the first calculations concerning the effect of an electric field on optical absorption associated with interband transitions, a complete understanding is far from achieved.

As most materials, especially in the region at the fundamental absorption edge, show interband transitions dominated by excitonic effects, and layer materials are no exception in this respect, theories based on the Franz-Keldysh effect are invalid. Fortunately, several calculations including the electron-hole interaction have appeared recently [5, 6], but these calculations are based on Elliot's [7] formalism of absorption by excitons, which assumes a spatially isotropic exciton and, at least in the case of layer materials, a quantitative comparison of theory and experiment is at present difficult. The anisotropy of the exciton parameters (dielectric constant and effective mass) in layer materials, makes Elliot's formalism unsuitable. In any case, broadening effects, other than those induced by the electric field, have not as yet been adequately dealt with theoretically.

Other problems that make quantitative comparison with theory difficult are the lack of accurate knowledge of the magnitude of the induced perturbation and, in EM for instance, the existence of surface states makes an accurate estimate of the strength of the applied field difficult to obtain without resorting to subsidiary experiments. Incomplete penetration of the applied electric field creates complications as well, which are linked to the spatial extension of the exciton states.

All these difficulties and how to overcome them will be presented in Section 5 together with a general model for electromodulation effects. Sections 2 and 3 are devoted to an outline of the currently available theories. The most commonly used techniques are described in Section 4, and from Section 6 onwards the experimental results on some layer materials and the proposed interpretations will be presented.

2. Optical Absorption by Solids

The foundations of the one-electron band theory of solids were laid by the early work of Bloch, Brillouin, Slater and Wigner. Towards the mid-1950's band structure calculations began to be related to optical properties. The basic principle that makes optical experiments a powerful tool in the study of electronic states is that absorption or emission of light is, at least beyond the fundamental absorption edge, accompanied by the transfer of electrons from one state to another in accordance with the necessary momentum and energy conservation rules. As a consequence, an absorption spectrum is, neglecting any variations in transition probabilities with photon energy, basically a display of the energy separation between electronic levels.

Many books and review articles on optical properties of solids are available and only a sketch of the most relevant facts to this work will be mentioned in this section.

Experiments have shown that some gross features of the optical spectra by solids can be interpreted within the scope of the one-electron band picture. In this approximation the effect of electro-magnetic radiation on the many-electron wave functions is obtained by calculating the response of the one-electron wave functions in a self-consistent potential.

If one considers only the interaction of the electrons with the electromagnetic field, it can be shown that the imaginary part of the dielectric constant (which is proportional to the absorption coefficient α divided by the photon energy $\hbar\omega$) is given by:

$$\varepsilon_2 \propto \int \frac{|\mathbf{e} \cdot \mathbf{M}_{vc}|^2 \, dS}{\nabla_k(E_c - E_v)}, \tag{1}$$

where
dS is an element of surface in **k**-space defined by $E_c - E_v = \hbar\omega$,
ω is the angular frequency of the radiation,
e is the polarization vector of light,
$\mathbf{M}_{vc}$ is the matrix element (transition probability) and is proportional to:

$$\int \psi_{kc}^*(\mathbf{r}) \, \nabla \psi_{kv}(\mathbf{r}) \, d\mathbf{r} \tag{2}$$

where $\psi_{kc}(\mathbf{r})$ and $\psi_{kv}(\mathbf{r})$ are the wavefunctions of the states in the empty conduction and filled valance band with energies E_c and E_v respectively.

A plot of ε_2 versus ω shows singularities at critical points where $\nabla_k E_c - \nabla_k E_v = 0$. This may happen because:

$$\nabla_k E_c(\mathbf{k}) = \nabla_k E_v(\mathbf{k}) \tag{3}$$

which may occur at any general value of the wave vector **k**, or because:

$$\nabla_k E_c(\mathbf{k}) = \nabla_k E_v(\mathbf{k}) = 0 \tag{4}$$

which occurs only at points of high symmetry in the Brillouin zone.

In the immediate vicinity of a critical point k_0 in the Brillouin zone, the energy $E(k)$ can be approximated by a quadratic function of $(\mathbf{k} - \mathbf{k}_0)$

$$E_c - E_v = E_0 + \frac{\hbar^2}{2} \frac{(k_1 - k_{01})^2}{\mu_1} + \frac{\hbar^2}{2} \frac{(k_2 - k_{02})^2}{\mu_2} + \frac{\hbar^2}{2} \frac{(k_3 - k_{03})^2}{\mu_3}, \tag{5}$$

where

$$\frac{1}{\mu_i} = \frac{1}{m_{ei}^*} + \frac{1}{m_{hi}^*}, \quad i = 1, 2, 3 \tag{6}$$

and m_{ei}^* and m_{hi}^* are the electron and hole effective masses in three principle crystallographic directions.

This generates four different types of critical points:
(a) M_0-type when $\mu_1\mu_2\mu_3$ are positive.
(b) M_1-type when $\mu_1\mu_2$ are positive; μ_3 negative.
(c) M_2-type when $\mu_1\mu_2$ are negative; μ_3 positive.
(d) M_3-type when $\mu_1\mu_2\mu_3$ are negative.

The shapes of these four types of critical points are shown in Figure 1.

One has to observe that momentum conservation requires that (neglecting the effects of phonons):

$$\mathbf{k}_v - \mathbf{k}_c = \boldsymbol{\eta} + \mathbf{G}, \tag{7}$$

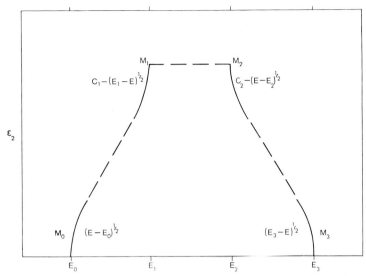

Fig. 1. Schematic representation of the four types of critical points according to the one-electron model theories.

where $\boldsymbol{\eta}$ is the wave vector of the electromagnetic radiation and $\mathbf{G}$ a reciprocal lattice vector. Now $\mathbf{k}_v - \mathbf{k}_c$ is of the order of 10^8 cm^{-1} while (in the region of the visible, for instance) $\boldsymbol{\eta}$ is the order of 10^4 cm^{-1} and consequently can be neglected.

One can also take $\mathbf{G} = 0$ (as $\mathbf{G}$ would be only important in processes like the so-called Umklapp enhancement. Consequently:

$$\mathbf{k}_c = \mathbf{k}_v \tag{8}$$

and we are concerned only with vertical transitions in the Brillouin zone. Such transitions are termed 'direct transitions'.

So far, only the interaction of the electrons with the electromagnetic field has been considered and from momentum conservation requirements it has been shown that only direct transitions are allowed. However, if one allows for the participation of phonons, then it is possible to have 'indirect transitions', where the momentum conservation rules are satisfied with the help of a phonon, created or absorbed during the process. As the main concern of this review is with direct transitions, indirect transitions will be left aside.

So far the most obvious of the many-body interactions has been neglected; this is the Coulomb-interaction between the photo-excited electron in the conduction band and the hole in the valence band. If one considers this interaction, there exists the possibility of forming bound electron-hole states, which receive the name of 'excitons'. One can look at excitons in two ways. An exciton can be considered as a packet of excitations made up of atomic excitations, that is, a linear combination of localized excitations, which implies that the electron making a transition does not leave the cell from which it originates. This sort of tightly-bound exciton is termed a FRENKEL exciton. On the other hand, one can look at an exciton as a conduction band electron and a valence band hole, bound together but possibly with a considerable separation, travelling through the crystal in a state of total wave vector K. This kind of exciton is called a WANNIER exciton.

Although equivalent in principle, the Frenkel and the Wannier excitons differ in that they have to be treated in different ways. In the former the electron sees the hole and the details of the lattice potential, while in the latter the electron sees the hole and the average lattice potential. Frenkel excitons appear, for instance, in molecular, rare gas and alkali halides crystals. Excitations of the Wannier type are found in semiconductors in which the bonding is normally more covalent than ionic (for references and details concerning the theory of excitons, see reference 8). The excitonic absorption in the majority of layer materials is of the Wannier type.

The theory for the optical absorption due to excitons at interband thresholds was worked out by Elliot [7], who showed that for direct transitions the imaginary part of the dielectric constant is given by:

$$\varepsilon_2(\omega) = \frac{4\pi e^2}{m^2 \omega^2} \frac{1}{V} \sum_{n,l,m} |\phi_{n,l,m}^{(0)}|^2 |\langle v |\mathbf{ep}| c \rangle|^2 \times \delta(\hbar\omega - E_n - E_g), \tag{9}$$

where $\phi_n(\mathbf{r})$ is the solution of the Schrödinger equation for a hydrogen-like atom:

$$\left\{-\frac{\hbar^2}{2\mu}\nabla^2 - \frac{e^2}{\varepsilon r}\right\} \phi_{n,l,m}(\mathbf{r}) = \left\{E_n - E_g - \frac{\hbar^2 k^2}{2M}\right\} \phi_{n,l,m}(\mathbf{r}) \tag{10}$$

where $\mathbf{r} = \mathbf{r}_e - \mathbf{r}_h$ is the coordinate of the electron relative to the hole.

$$\mu = \frac{m_e^* m_h^*}{m_e^* + m_k^*} \quad \text{and} \quad M = m_e^* + m_h^*, \mathbf{K} = \mathbf{K}_e + \mathbf{K}_h, \tag{11}$$

where $\langle v| \mathbf{ep} |c \rangle$ is written for $e \int \psi_{kc}^*(\mathbf{r}) \cdot \nabla \cdot \psi_{kv}(\mathbf{r}) \, d\mathbf{r}$.

From (9) one can see that the probability of creating an exciton is proportional to the probability of finding the electron and the hole at the same point in space. The momentum conservation rules require that $K_e = -K_h$, so the only transitions that can take place are effectively at $K = 0$, and that is the reason for the name of 'direct excitons'. The solution of the effective mass Schrödinger Equation (10) leads to a series of discrete states for $E < E_g$ and a continuum for $E > E_g$.

In the discrete set the energies are given by:

$$E_n = E_g - \frac{\mu e^4}{2\hbar^2 \varepsilon^2 n^2}, \tag{12}$$

where n is the principle quantum number for the hydrogen-like atom and $\phi(0)$ will be non-zero only for S-states, for which:

$$|\phi_n(0)|^2 = (\pi a_0^3 n^3)^{-1}, \tag{13}$$

where a_0 is the effective Bohr radius for the exciton ground state and is given by:

$$a_0 = \frac{\varepsilon \hbar^2}{\mu e^2} \tag{14}$$

From (13) one can see that the intensity of the lines will fall like n^{-3} and, from the energy equation (12), that their separation decreases with increasing n. When $n = \infty$, the continuum is reached at E_g.

Experimentally, exciton absorption never leads to delta function lines. Many reasons for broadening exist such as scattering by impurities and defects, size effects, temperature, etc. Temperature broadening is one of the main effects involved in producing a finite width for an exciton line. No general satisfactory theory has yet been worked out. The line shape for exciton absorption expected where temperature effects are included has been calculated [9] for the limiting cases of strong exciton-phonon coupling and weak exciton-phonon coupling leading to Lorentzian and Gaussian shapes respectively. Nevertheless, broadening has been included as a parameter Γ and reasonable predictions of line shapes can be obtained in this way. The theoretical absorption spectrum for an M_0-edge according to Elliott's theory is shown for various broadening parameters of the exciton discrete states [21], in Figure 2.

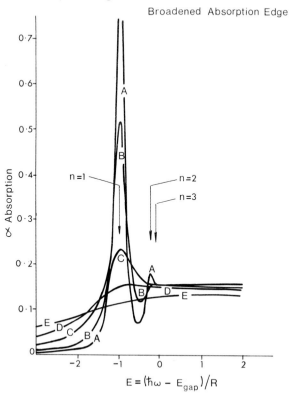

Fig. 2. Broadened M_0-edge according to Elliot's theory of excitons for several values of the broadening parameters: A, $\Gamma/R = 0.1$; B, $\Gamma/R = 0.2$; C, $\Gamma/R = 0.5$; D, $\Gamma/R = 1.0$; E, $\Gamma/R = 2.0$.

So far the theory is valid for isotropic excitons, a treatment due to Deverin [11], gives an idea of how anisotropy will affect the exciton states. Confining oneself to the case of uniaxial crystals which is most common among layer materials one has to account for the anisotropy by introducing effective masses and dielectric constants of a tensorial nature. If one defines a Z-direction parallel to the c-vector axis one has:

$$\mu_x = \mu_y = \mu_\perp, \qquad \varepsilon_x = \varepsilon_y = \varepsilon_\perp$$
$$\mu_z = \mu_\parallel, \qquad \varepsilon_z = \varepsilon_\parallel \tag{15}$$

with which one can define an anisotropy parameter

$$\gamma = \frac{\mu_\perp \varepsilon_\perp}{\mu_\parallel \varepsilon_\parallel} \tag{16}$$

Introducing a new system of coordinates

$$\rho = (\xi, \eta, \zeta) = \left(x, y, \sqrt{\frac{\mu_\perp}{\mu_\parallel}} z\right) \tag{17}$$

Equation (10) can be re-written in the form:

$$\left\{\frac{\hbar^2}{2\mu_\perp}\left(-\frac{\partial^2}{\partial \xi^2} - \frac{\partial^2}{\partial \eta^2} - \frac{\partial^2}{\partial \zeta^2}\right) - \frac{e^2}{\sqrt{\varepsilon_\perp \varepsilon_\parallel}}\frac{1}{\sqrt{\xi^2 + \eta^2 + \zeta^2}} + P(\gamma)\right\}\phi(\mathbf{p})$$

$$= E_{\text{exc}}\phi(\rho), \tag{18}$$

where

$$P(\gamma) = \frac{e^2}{\sqrt{\varepsilon_\perp \varepsilon_\parallel}}\left\{\frac{1}{\sqrt{\xi^2 + \eta^2 + \zeta^2}} - \frac{1}{\sqrt{\xi^2 + \eta^2 + \gamma\zeta^2}}\right\}. \tag{19}$$

If one treats $P(\gamma)$ as a perturbation, the zero order solutions of (18) have eigenvalues given by:

$$E^n_{0_{\text{exc}}} = -\frac{\mu_\perp e^4}{2\varepsilon_\perp \varepsilon_\parallel \hbar^2}\frac{1}{n^2}, \qquad n = 1, 2, \ldots, \infty \tag{20}$$

and localized within radii:

$$a^\mu_{0_{\text{exc}}} = \frac{\hbar^2 \sqrt{\varepsilon_\perp \varepsilon_\parallel}}{\mu_\perp e^2} n^2, \qquad n = 1, 2, \ldots, \infty. \tag{21}$$

From (20) one can see that despite the anisotropy one can still classify the zero order states in terms of the quantum numbers n, l, m of the hydrogen levels. The first order energy corrections given by perturbation theory are:

$$\langle n, l, m | P(\gamma) | n'l'm' \rangle = \int \phi^*_{n,l,m} P(\gamma) \phi_{n,l,m} \, d^3\mathbf{p} \tag{22}$$

as $P(\gamma)$ reduces the symmetry of the system it will consequently remove some of the degeneracies. $P(\gamma)$ belongs to $D\infty h$ (rotation about the c-axis leave the system invariant), consequently one can invoke group theory to find out which matrix elements (22) will be non-zero. Not only that, but matrix elements (22) will couple states of adequate symmetry leading to a mixing of orbitals.

Deverin [11] discussed (22) and the results are that $P(\gamma)$ is diagonal for the $n = 1, 2$ discrete states as $P(\gamma)$ does not couple s and p states but from $n = 3$, $P(\gamma)$ couples the states $3s$ and $3d_0$ leading to hybrid states. Deverin's findings are summarized in Figure 3. For the $n = 1$ state, the effect of the anistropy is to shift the energy position away from the exciton continuum for $\gamma < 1$, effectively increasing its binding energy, and towards the continuum for $\gamma > 1$. A similar effect can be seen for the $n = 2$ state except that in this case its degeneracies are lifted and the exciton state splits into three levels. For the $n = 3$ state the degeneracies are lifted as well, but the anisotropy couples the $3s$ and $3d_0$ states leading to the formation of the $|3-0\rangle$ and $|3+0\rangle$ states. For a detailed discussion the reader is referred to ref. 11. As it will be shown in Section 7 the effects of anisotropy on exciton formation can be usefully studied with the use of modulation techniques.

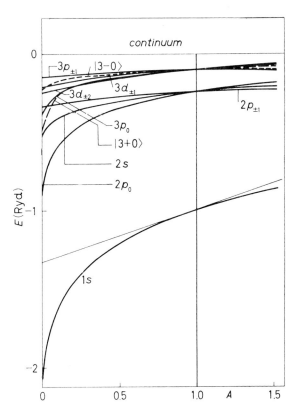

Fig. 3. The effect of the anisotropy on the first three exciton bound states.

3. On the Theory of Electromodulation

The theoretical groundwork for electromodulation spectroscopy was laid by Franz [3] and Keldysh [4] who independentlly performed the first calculations concerning the effect of a strong electric field on optical absorption associated with interband transitions.

The Franz-Keldysh effect was later extended [12, 13, 14], always in the framework of a one-electron approximation. In the case of an M_0-edge the primary prediction of all these theories is that the optical absorption edge broadens and shifts towards lower energies and in addition electric field induced oscillations appear above the edge.

Physically one can understand the Franz-Keldysh effect as follows: In real space, the effect of an electric field is to tilt the band edges, which remain parallel and separated vertically by the energy gap. The wave functions of the conduction and valence band states will have exponentially decaying tails extending into the forbidden energy gap (the potential energy would be negative, or if one defines 'momentum' by $p^2/2m^* - eFz = E$, this corresponds to imaginary momentum within the gap, leading to an exponentially decaying wave function). A schematic illustration is given in Figure 4(a) and (b). In a plot of energy versus position, the local nature of the electro-magnetic perturbation requires the transition to be

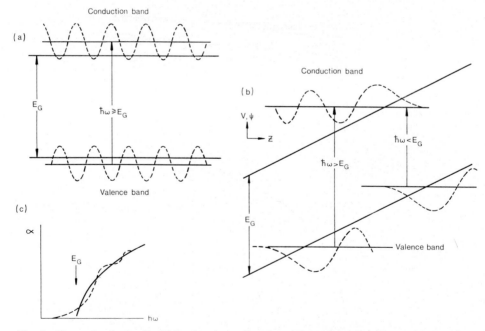

Fig. 4. Energy band edges (solid lines) and wave functions (dashed lines) in real space of the valence and conduction bands (a) without an applied electric field, (b) with an applied electric field along the z direction, (c) absorption coefficient for an M_0-edge without an electric field (solid line) and with an electric field (dashed line in the one-electron model approach.

vertical and, in the regions where there are exponentially decaying tails in the forbidden energy gap, it is possible to have photon induced transitions with energy less than the gap. However, the oscillator strength here is much smaller than it is for the higher energy band-to-band transition. The absorption coefficient for an M_0-edge without field is shown by the full line in Figure 4(c), while the broken line corresponds to the absorption coefficient when a field F is applied. The theoretically predicted field-induced changes in ε_2 for M_0, M_1, M_2 and M_3 edges are shown in Figure 5 [15].

Although some authors claim that the Franz-Keldysh effect can account for the many experimentally measured electromodulation spectra in the limit of high fields and high temperatures [16] the importance of excitonic effects has been known for a long time [15, 17]. One-electron theories do not include the most obvious of the many-body effects. This is the Coulomb interaction between an electron in the conduction band and the hole left behind in the valence band, which as mentioned in Section 2, leads to exciton formation, due to the creation of bound electron-hole pairs.

An electric field applied along the z-axis, for instance, will only affect $|\psi_n(0)|^2$ in Equation (9). Then the problem of finding the effect of an electric field on the optical absorption by excitons is reduced to solving the Schrödinger equation for a hydrogen-like atom in the presence of an electric field. The exact analytical solution of this equation has not yet been obtained and theoreticians have been forced to use approximate models [18, 19] or numerical integrations [20, 5, 6].

A criticism commonly made to all these theories is that none of them incorporate lifetime broadening of the excitons, but treat, for zero applied field, the exciton discrete states as delta functions, (see formula 9). But in fact the broadening induced by the electric field is usually much smaller than the lifetime broadening. Broadening has been introduced as a mathematical parameter in order to fit the experimental data [10, 21] and, although very good fits have been obtained, broadening has never been incorporated in physically meaningful ways.

Before summarizing the main theoretical predictions for electric field effects on excitons, a look at the physics of the problem should give some qualitative understanding of the effect.

In Figure 6 the potential energy seen by the electron under an applied electric field, as a function of its coordinates relative to the hole, is illustrated by the full line (the electric field being applied along the z-axis). For an electronic s-like state with energy E below the classical ionization energy, one can qualitatively predict its wave function $\psi(z)$. Between points $-z_1$ and $-z_3$ the state will find a potential barrier, (a region of imaginary 'momentum'), and the maximum of the potential barrier will be at $z_2 = -(|e|/eF)^{1/2}$. The electronic state will meet another potential barrier for value of z bigger than z_4. The values of z_1, z_3 and z_4 will be given by the solution of $(e^2/\varepsilon z) \pm eFz = E$. As a result, $\psi(z)$ will oscillate for values of $z < -z_1$, be exponentially damped up to the maximum of the potential barrier $-z_2$, rise exponentially up to $-z_3$, be fairly flat for an s-like state between

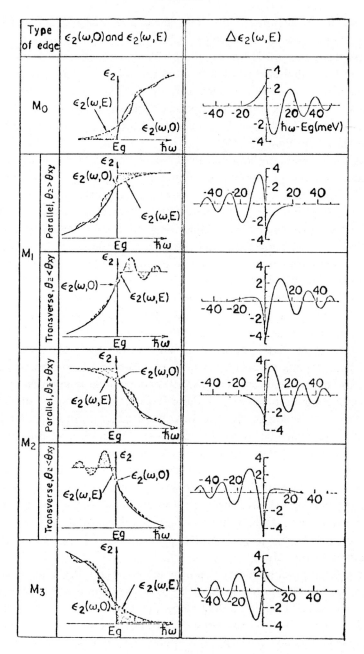

Fig. 5. The theoretically predicted electric field induced changes in the imaginary part of the dielectric constant for the several types of edges expected from one-electron models.

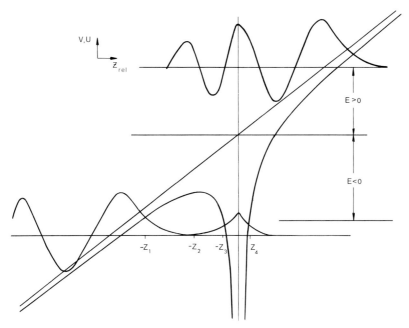

Fig. 6. Potential and wave functions for the relative motion of electron and hole as a function of the electron-hole separation, $z = z_e - z_h$, for an electric field applied along the z direction.

$-z_3$ and z_4 and be exponentially damped for $z > z_4$. Comparing with Figure 4 one can see that inclusion of the electron-hole interactions means that $\psi(z)$ acquires an exponential rise as z approaches zero and is totally undamped near the origin. Consequently one should expect the optical absorption coefficient to be exponentially enhanced.

Looking at the problem from a classical point of view, a criterion for ionization of the nth exciton state (with zero-field binding energy $-R/n^2$) is when a potential energy drop of one exciton Rydberg across the effective Bohr radius is induced by the electric field. This ionization field is given by $F_I(n) = Rn/ean = (\mu/m)^2 \times \varepsilon^{-3} \times n^{-4} \times 2.59 \times 10^9$ V cm^{-1} for the nth state. Application of an electric field $F = F_I(n=1)$ would mean that the exciton ground state lies above the potential barrier at z_2 in Figure 6. In general the exciton ground state can be ionized for applied fields $F < F_I$ because of quantum mechanical tunnelling effects.

The ionization field $F_I = (\mu/m)^2 \varepsilon^{-3} \times 2.59 \times 10^9$ V cm^{-1} is used as the electric field unit in the theoretical calculations and it will be used extensively in this work.

The main conclusion reached by the numerical calculations are:

(i) Well below the edge, the logarithm of the absorption coefficient varies very nearly linearly with photon energy. Figure 7(a) shows the shape of the calculated absorption edges for different fields in units of F_I, while Figure 7(b) corresponds to the Franz-Keldysh effect [6].

(ii) The field dependence of the energy position of the exciton ground state is

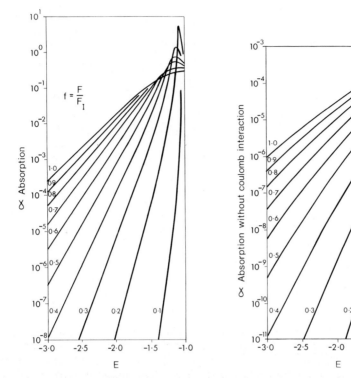

Fig. 7. The calculated effect of an electric field on an M_0-edge for several values of the applied electric field which is measured in units of the exciton ionization field F_I. (a) the electron-hole interaction is included, (b) the electron-hole interaction is excluded. The horizontal axis is measured in units of the exciton Rydberg from the exciton continuum.

given in Figure 8 [5]. The field is measured in units of F_I and the energy in units of the exciton Rydberg. For fields $F < 0.85 F_I$ the exciton ground state undergoes a shift to lower energies, in the manner predicted by perturbation theory (Stark effect) but for $F > 0.85 F_I$ it moves towards higher energies (the ground state begins to tunnel out of the potential well).

(iii) Other bound states $n = 2, 3, \ldots$ in addition to being shifted and broadened in a similar manner, undergo a splitting according to their degeneracies. Because of their lower binding energies, these higher order states will, for the same field, suffer relatively larger effects. Several theoretical excitonic EA spectra, representing the change in the imaginary part of the dielectric constant, $\Delta \varepsilon_2$, and for several broadening parameters Γ in units of exciton Rydbergs are shown in Figure 9. The temperature and field dependences of the peak heights and energy separations of the electromodulated response are shown in Figure 10 and Figure 11 [5].

(iv) The height of the satellite peaks h_1 and h_3 in $\Delta \varepsilon_2$ arise from the quenching and broadening of the exciton ground state by the field; h_1 is greater than h_3 for small fields and broadenings [5].

(v) h_1 and h_3 can increase or decrease with the applied field (Figure 10) but h_2,

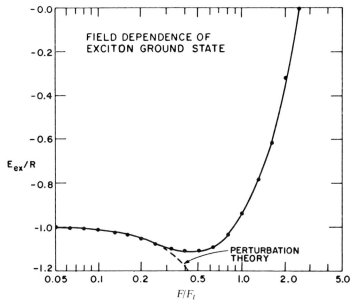

Fig. 8. Field dependence of the exciton ground state energy position for a hydrogenic model. The position is measured in units of the exciton Rydberg from the exciton continuum and the electric field is measured in units of the exciton ionization field F_I.

the main peak in $\Delta\varepsilon_2$, should never decrease with field; for large fields h_2 approaches a constant value.

(vi) For values of $(F/F_I)5$ the position of the first negative peak occurs at $E_g - R$ (see Figure 11) [5].

(vii) The energy separation of the zero line crossing points associated with the second oscillation $\Delta\varepsilon_2$ increases as $(F/F_I)^{2/3}$, independent of temperature as shown in Figure [5].

These are the main theoretical predictions that in principle should be checked with experimental data. However any comparison has to allow for the existence of experimental complications (internal fields, incomplete field penetration, thickness effects, etc....). These and any fundamental departures from the ideal case, in particular sample anisotropy, should be incorporated for a proper quantitative analysis.

In the case of layer materials it is obvious that the anisotropy of the lattice will have to show somehow on the electromodulation spectra. Recently calculations on the effect of electric fields on the exciton ground state energy position in an anisotropic environment have been made [22] using a variational method. Unfortunately this method only yields exact results for small values of the applied electric field, that is in the region of the Stark effect. For an isotropic exciton the quadratic Stark shift of the exciton ground state has a Stark coefficient $r = \frac{9}{4}$ ($\Delta E = r |F|^2$). The effect of the anisotropy on the Stark shift is illustrated in Figure 12 for anisotropy parameters smaller than unity. One can see that the Stark

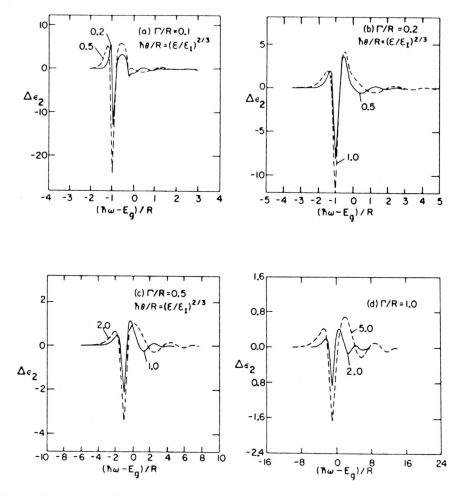

Fig. 9. Temperature and electric field dependence of ΔE_z as a function of $(E-E_g)/R$ for (a) $\Gamma/R = 0.1$, 0.2 and 0.5; (b) $\Gamma/R = 0.2$, $F/F_I = 0.5$ and 1.0; (c) $\Gamma/R = 0.5$, $F/F_I = 1.0$ and 2.0; and (d) $\Gamma/R = 1.0$, $F/F_I = 2.0$ and 5.0.

coefficient r becomes smaller for decreasing anisotropy parameters and that its value depends on the orientation of the electric field. The Stark coefficient is predicted to be smaller for fields perpendicular to the c-axis than for fields parallel to the c-axis which can be physically understood from the fact that the exciton is more extended along the layers than perpendicular to them. These calculations suggest that electromodulation can be used as a technique to study the anisotropy of exciton parameters, and, although they break down in the region of high fields, an extrapolation of the results for the high field region leads to good estimates of the parameters of the exciton at the fundamental absorption edge of PbI_2 [23]. This is discussed in detail in Section 7.

When $A > 1$, the Stark coefficient becomes bigger than in the isotropic case

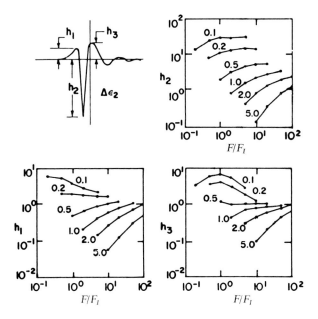

Fig. 10. Temperature and electric field dependence at amplitudes of the field induced change in the imaginary part of the dielectric current for $\Gamma/R = 0.1, 0.2, 0.5, 1.0, 2.0$ and 5.0.

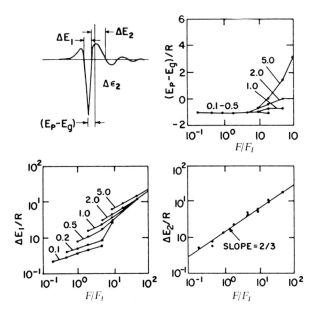

Fig. 11. Temperature and electric field dependence of energy differences in the field induced change in the imaginary part of the dielectric constant for $\Gamma/R = 0.1, 0.2, 0.5, 1.0, 2.0$ and 5.0.

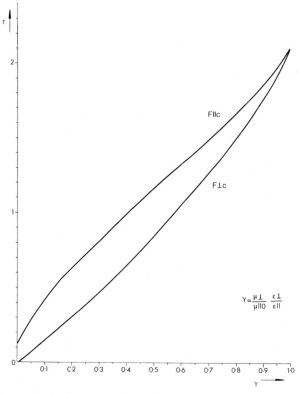

Fig. 12. Effect of the anisotropy ($\gamma < 1$) on the Stark coefficient of an exciton in a layer material for the electric field applied parallel and perpendicular to the c-axis.

and again it reflects the anisotropy of the lattice by its dependence on the orientation of the electric field.

Exact calculations in the high field regions are called for at present.

Another approach to the theory of EM has been recently proposed [24]. When an electric field is applied to a solid, the electrons will be accelerated and consequently momentum is no longer a good quantum number (in other words the term $eF \cdot z$ destroys the translational invariance of the Hamiltonian in the direction of the electric field). As a result the one-electron Block functions that describe the unperturbed electronic states of the crystal will become mixed. This is equivalent to spread the sharp vertical transition which occur at a critical point of the unperturbed K-space over a finite range of initial and final states. For small values of the electric field it turns out that the electric field induced charge in the dielectric constant is given essentially by the third derivative of the unperturbed dielectric constant. Excitonic effects can be included in the theory if the electron-hole interaction is represented by means of a Koster-Slater model potential [25]. Nevertheless the excitonic effects are not introduced satisfactorily, as the Koster-Slater potential can account at most for one bound state. Figure 13 gives a visual display of the idea behind the principle of third derivative spectroscopy.

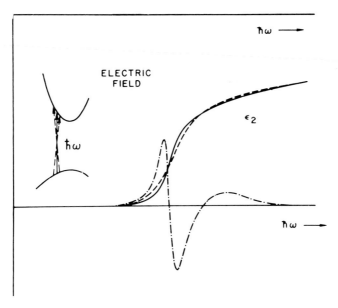

Fig. 13. Illustration of the idea behind the principle of third derivative spectroscopy. Because of the effect of the applied electric field, transitions other than vertical will be allowed leading to a third derivative like spectrum.

4. Experimental Techniques of Electromodulation

In any modulation experiment one has to consider two experimental problems. First is how to apply the modulated perturbation to the sample and second is how to detect the change in the transmitted or reflected light.

For electromodulation experiments three techniques have been used in order to apply the modulating electric field, and they receive the name of electrolyte, transverse and longitudinal method, respectively.

The electrolyte method involves modulation of the space-charge layer at a semiconductor-electrolyte interface. For a detailed discussion of the details of this experimental arrangement the reader is referred to ref. 1. The main criticism to this technique is that the experimentalist is limited to work at temperature above the freezing point of the electrolyte, rendering low temperature work impossible to perform. In addition one has to avoid surface contamination of the sample from the electrolyte.

For materials with high resistivity, typically bigger than $10^8 \, \Omega \, cm^{-1}$, the transverse method can be used. This method consists of having two electrodes in a planar or gap configuration on the surface of the material. Evaporated metallic electrodes [26, 27] or slip-on metal caps [28] have been used. The typical experimental arrangement is shown in Figure 14. The electrode gap can be easily achieved by placing a thin flat galvanometer wire on top of the crystal. When the width of the gap between electrodes is much bigger than the thickness of the

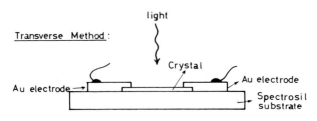

Fig. 14. Schematic diagram of a typical experimental arrangement for the transverse method.

sample the electric field can be considered uniform all the way through the sample [29]. However for layer materials this geometrical condition may be more stringent than for isotropic crystals, and it is desirable to take into account the difference in the conductivity perpendicular and parallel to the layers.

One of the obvious advantages of the transverse configuration is that one has a surface of the crystal absolutely free of contacts making the arrangement very suitable for optical experiments. In addition one can vary the orientation of the field with respect to the crystallographic axis and the polarization vector of the incident light. This has proved useful in the study of anisotropy (see, for instance, Section 19 on As_2Se_3). The transverse configuration also allows the sample to be easily cooled to very low temperatures. All these considerations make this method the most desirable. However if the material has a resistivity smaller than $10^8 \, \Omega \, \text{cm}^{-1}$, the current passed through the sample is high enough to produce thermomodulation effects which can mask the electromodulation response.

To study electromodulation at low temperatures on materials with low resistivities one is forced to use the longitudinal or 'package' method based on the 'field effect' [2]. In this arrangement the electric field is applied to the crystal via an insulator. The latter is used to avoid injection of carriers, which would produce unwanted heating and other undesirable effects.

A schematic diagram of a typical sample arrangement is shown in Figure 15. A package that has proved very successful [30], when dealing with very small crystals, consists of a substrate of Spectrosil coated with a transparent electrode of SnO_2 (Nesa glass) on top of which a thin crystal is laid. An insulating layer of 12 μm of Mylar is attached to the crystal with a thin coating of 'Formvar' and on top of this a semitransparent electrode of gold is evaporated. For an experimental arrangement of this kind and under the assumptions that no surface states are

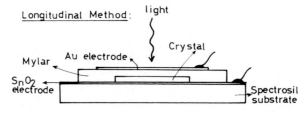

Fig. 15. Schematic diagram of a typical experimental arrangement for the longitudinal method.

present one can estimate the field at the surface of the sample (but inside the sample) to be given by:

$$F = \frac{\varepsilon_I}{\varepsilon_{xtal}} \frac{V_{applied}}{t_I}, \qquad (23)$$

where ε_I and ε_{xtal} are the dielectric constants of the insulating layer and the crystal respectively and t_I the thickness of the insulator. Equation (23) is only valid when no surface states exist at the interface. If surface states are present the field is reduced by an amount proportional to the charge per unit area residing in surface states (Q_{ss}) and the field inside the sample at the surface is given by:

$$F = \frac{V_{applied}}{t_I} - 4\pi Q_{ss} \frac{\varepsilon_I}{\varepsilon_{xtal}}. \qquad (24)$$

Without subsidiary experiments Q_{ss} is normally unknown. Although this complication is recognized no correction is normally made for it.

A conventional energy band diagram of the longitudinal arrangement is shown in Figure 16. The electrostatic potential inside the sample can be found by integrating Poisson's equation:

$$\frac{d^2 V}{dx^2} = \frac{4\pi \rho(x)}{\varepsilon_{xtal}}, \qquad (25)$$

where x is the coordinate from the surface of the sample towards the inside of the sample and $\rho(x)$ is the local charge density. Solving Equation (25) under different situations gives rise to different voltage configurations and different penetrations of the electric field. Consequences of this will be discussed in later sections (see Section 5 and the section on the transition metal dichalcogenides).

Obvious disadvantages of the longitudinal arrangement are the difficulty in

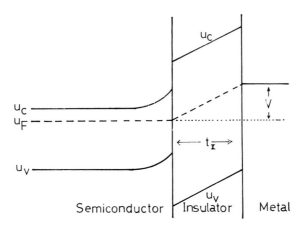

Fig. 16. A conventional energy band diagram of the longitudinal arrangement.

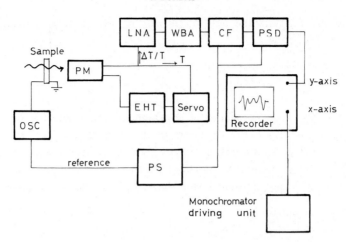

Fig. 17. A typical electronics detecting system for electromodulation measurements.

obtaining accurate estimates of the electric field and accounting for its non-uniformity. However it does have one advantage over the transverse configuration in that higher fields can be applied before breakdown is reached.

The detection of the induced charge in the reflectivity or transmission of the sample is pretty straightforward. A typical arrangement [30] is shown in Figure 17. The photomultiplier (PM) yields two signals, in this case ΔT and T. T is fed into a servo control of the high voltage power supply (EHT) which keeps T constant, while $\Delta T/T$ is fed into a low noise amplifier (LNA), wide band amplifier (WBA), coherent filter (CF) and a phase sensitive detector (PSD). The CF and the PSD are driven by a reference signal, which is provided by the oscillator (OSC) which is used to apply the electric field to the sample. The reference is phase shifted (PS) before entering the CF and the PSD. The output of the PSD drives the y-axis of a recorder, which has its x-axis driven by a signal coming from the driving units of the monochromator.

Many different variations of this electronic system can be used, but essentially they all have the same basis.

5. A General Model for Electromodulation Signals

Very often in the course of an electromodulation experiment one can detect signals at the frequency of the applied electric field and at the first harmonic. In addition different shapes for these signals can be obtained for electric fields of different polarities.

In this section a model which accounts for most of the effects found in electromodulation experiments performed on the excitons at the fundamental absorption edge is presented and discussed. This model will be of particular importance for explaining the EM results on the group VI transition metal dichalcogenides.

The following symbolism will be used:

F_a: The applied alternating electric field
F_A: The peak-to-peak value of F_a
F_b: Any applied d.c. bias
F_e: The effective field seen by the exciton
F_i: Any internal field present in the absence of an applied one
f: Frequency of the applied field
$2f$: Frequency of the 1st harmonic of the applied field
S_f: Signal detected at f
S_{2f}: Signal detected at $2f$
$\Delta\alpha$: $\alpha(F_1) - \alpha(F_2)$: The change in the optical absorption coefficient when two different fields are applied

As a definition, the sign of $\Delta\alpha$ will be taken as positive (negative) when the observed response is in phase (out of phase) with the applied alternating field. This definition is appropriate only when referring to S_f: for S_{2f} there is an ambiguity.

For those electromodulation experiments in which the electric field is applied via an insulator in the longitudinal geometry, F_a refers to the field at the surface of the crystal but inside the crystal, and is calculated from the applied voltage by Equation (23). This equation assumes that the density of electronic states at the interface between the insulator and the crystal is sufficiently low to produce negligible screening which would otherwise reduce the applied field.

A starting point in the interpretation of any electromodulation spectrum is that an exciton cannot distinguish between a positive and a negative electric field: the effect of both should be exactly the same. Consequently in the absence of screening effects coming from electronic states at the interface, the field seen by the exciton is given by:

$$F_e = |F_a + F_b + F_i|. \tag{26}$$

However, in the more conducting crystals the electric field penetration may not be uniform. The effects of such incomplete field penetration will be discussed in this section under Case A and in the sections on those materials in which they are important.

Under the assumption of no screening by surface states and uniform electric fields (except where mentioned specifically), several situations will now be discussed.

CASE A

A simplified situation, often met, is one in which the internal field is zero, $F_i = 0$, $F_e = |F_a + F_b|$. Now if the applied field is sinusoidal about zero, the negative and positive half cycles are equivalent as far as any effect on the exciton is concerned and F_e has the shape shown in Figure 18 A_1. As F_e has no Fourier component at f, S_f will be zero and the modulated response will be detected at $2f$. To a 1st approximation, $\Delta\alpha = \alpha(F_A/2) - \alpha(0)$ although in order to be more exact one

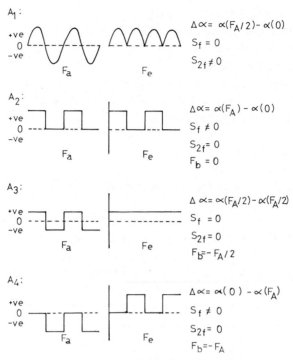

Fig. 18. Illustration of the effect of an applied electric field in the absence of internal fields on the optical absorption by excitons.

should average the time varying fields (an unnecessary complication that can be avoided with the use of square wave modulating voltage, as shown below).

With the use of square wave modulation oscillating from zero to a positive value (as F_a in Figure 18 A_2), F_e follows F_a. As a square wave has no first harmonic, $S_{2f} = 0$, and the signal will be detected at $f(S_f \neq 0)$. Neglecting any deformation of the square wave by sample and lead capacitance (which in practice is always present to some extent) $\Delta\alpha = \alpha(F_A) - \alpha(0)$.

If one biases the square wave so as to have equal positive and negative excursions ($F_b = -(F_A/2)$) (Figure 18 A_3) then F_e will be constant and no modulation will be achieved $\Delta\alpha = \alpha(F_A/2) - \alpha(F_A/2) = 0$.

If further biasing is applied so as to have $F_b = -F_A$ then F_e will suffer a change of phase with respect to F_a and a change of sign in $\Delta\alpha$ will occur relative to case A_2 ($\Delta\alpha = \alpha(0) - \alpha(F_A)$). Then $S_f \neq 0$ and $S_{2f} = 0$ because of the same arguments applicable to case A_2.

If even further biasing is applied so as to have $F_b < -F_A$ then $\Delta\alpha = \alpha(F_b + F_A) - \alpha(F_b)$. In general this situation is not experimentally desirable as usually one is interested in having modulation from the zero field position (as in A_2 and A_4) whenever possible.

In order to avoid confusion some attention should be given to what is meant by a 'change of sign' in $\Delta\alpha$.

If one had achieved the change of the polarity of the applied field between situation A_2 and A_4 not by biasing but by throwing a switch on the pulse generator so that the applied voltage swung negative in the same cycle that it had previously been positive, *but the phase of the reference voltage from the generator to the phase sensitive detector had remained unchanged*, then there would not have been any change of sign of the signal as seen on the recorder. But to conform to our definition of sign, we would have had to call the second signal negative as the phase of the applied field would have been changed by π.

Although situations A_2 and A_4 have, *apart from the change of sign*, been shown to give equivalent signals, there are physical reasons why they can (and in fact sometimes do) give different shapes and magnitudes for $\Delta\alpha$.

Three circumstances are envisaged when this situation can arise. (1) Presence of surface states; (2) Incomplete penetration of the applied field; (3) Presence of internal fields.

The effect of the surface states can be illustrated by a simple example. Suppose that the net charge in the surface states is zero and there is thus no internal field due to space charge. The distribution of surface states with respect to the Fermi level however need not be symmetrical and their screening effect could conceivably be different for one direction of the applied field than the other. In a limiting case F_e could be equal to F_a for one polarity and zero for the other.

One should recognize as well the possibility that the response time of the surface states (for changing their charge state) may be so slow that they are effective in screening the applied d.c. bias but ineffective as far as a.c. signals are concerned. Thus in electromodulation experiments the effect of d.c. biasing may be zero and applied fields such as shown in A_2, A_3 and A_4 may thus lead to identical signals. For no internal fields this signal would reach zero (as in Case A_3) after a time sufficiently long that such slow states have reached their equilibrium condition.

This point will be referred to again in Case B.

The second possibility arises from incomplete penetration of the applied fields, which can lead to an asymmetry in the observed signal with reversal of polarity in the case of essentially unipolar conduction, such as would be expected if the material were highly *n*-type for example. Then application of positive and negative voltages to the electrode would lead to accumulation and depletion layers respectively. The latter will penetrate to a greater depth than the former and therefore will have a larger effect on the optical transmission. An effect of this kind is believed to occur for instance in $2H$—MoS_2 and WS_2 and its consequences are discussed in Section 12. The effect of internal fields for various conditions of biasing will be discussed in the next section.

Case B

Leaving aside, for the moment, screening by surface states and incomplete penetration, the only difference between this case and Case A is that here we

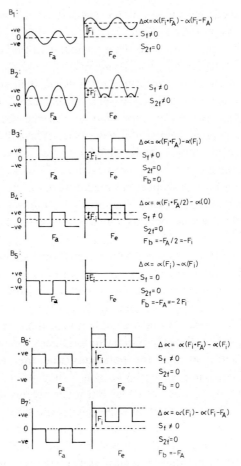

Fig. 19. Illustration of the effect of an applied electric field in the presence of an internal electric field on the optical absorption by excitons.

assume there is an internal field F_i, plus about the same order of magnitude as F_A. The origin of F_i could arise from contact potential differences, or for other reasons but these will not be discussed further here.

If F_A is sinusoidal and $F_A \leq F_i$ we have the situation displayed in Figure 19 B_1. In this case F_e will have no first harmonic and consequently the electromodulation response will be detected at f ($S_f \neq$ and $S_{2f} = 0$).

A rather more complicated situation appears when F_a is sinusoidal and F_A F_i. F_e will then have the shape shown in Figure 19 B_2. If one performs a Fourier analysis, F_e is found to have the following analytical form:

$$F_e = \frac{2}{\pi}\sqrt{1-\left(\frac{F_i}{F_A}\right)^2} \cdot F_A + \frac{4t_0}{T}F_i - \left\{\frac{4}{\pi}\sqrt{1-\left(\frac{F_i}{F_A}\right)^2} \cdot F_i - \frac{4t_0}{T}F_A\right\}$$
$$\times \sin \omega t - \left\{\frac{4}{\pi}\sqrt{1-\left(\frac{F_i}{F_A}\right)^2} \cdot F_A - \frac{8}{\pi}\left(1-\left(\frac{F_i}{F_A}\right)^2\right)^{3/2} F_A\right\} \cos 2\omega t, \quad (27)$$

where T is the period of F_a and t_0 is given by:

$$\cos \omega t_0 = \sqrt{1 - \left(\frac{F_i}{F_A}\right)^2}. \tag{28}$$

It can be seen that in this case $S_f \neq 0$ and $S_{2f} \neq 0$ and in principle it should be possible to determine the value of F_i from the known value of F_A and the detected S_f and S_{2f}.

However, using sine wave modulation in order to accomplish this would be a rather unnecessary complication as the situation becomes much simpler with the use of square wave modulations.

If F_a is a square wave having positive excursions and there is a built-in F_i, then F_e will have the form shown in B_3; Figure 19. As a square wave has no 1st harmonic, $S_{2f} = 0$ and the modulated response will be detected at f.

If one biases F_a so as to have one half cycle going positive and the other half going negative, F_e will be such that it will tend to reach the zero field position and if $F_A \geq F_i$ it is possible to achieve modulation from the zero field position. The special case of $F_A = 2F_i$ is illustrated in B_4 Figure 19. If further biasing is applied, S_f will begin to decrease in size for a constant F_A and it should be possible to achieve a total disappearance of S_f when $F_b = -2F_i$. This case is illustrated in B_5 Figure 19.

If still further biasing is applied S_f would appear again, having suffered an apparent change of sign with respect to Case B_3 Figure 19. This case is not illustrated.

A further case needs to be considered, that is $F_i > F_A$. For F_a going from zero to a positive value, $\Delta \alpha = \alpha(F_i + F_A) - \alpha(F_i)$, as illustrated in case B_6 Figure 19. For F_a going from zero to a negative value $\Delta \alpha = \alpha(F_i) - \alpha(F_i - F_A)$, as illustrated in Case B_7 Figure 19. If F_i is significantly larger than F_A, then there will be very little change in the observed signal, i.e. little change of signal with bias, between these two cases. This effect whenever found should be taken as an indication of the possible existence of a very high internal field F_i.

As for Cases A (no internal field) asymmetrical distribution of surface states would influence the responses for Cases B and in Case B_5; for example a finite signal S_f could arise. In addition we again consider the possibility of complete screening of the applied d.c. bias by surface states leading to identical signals for Cases B_3, B_4 and B_5 which, after a time sufficiently long that such slow states have reached their equilibrium condition, would be that corresponding to B_4.

Finally the effects of incomplete penetration of the applied field should be considered for Case B analogous to Case A.

6. Electromodulation Experiments on Layer Materials

When some conclusions of general validity are tried to be drawn from the published EM data, one is confronted with a clear cut division. On the one hand there are those experiments concerned with the region of the fundamental absorption edge, where the consideration of excitonic effects is of paramount importance in order to interpret the experimental data. On the other hand there are those experiments carried out in the photon energy region well above the fundamental absorption edge, where it is sometimes claimed that excitonic effects are not of the same importance, although it is often recognized that their inclusion is necessary if an understanding of the data is to be achieved.

For the sake of clarity it seems natural to divide the account of the EM data between the experiments at the fundamental absorption edge and the rest. In the former region, usually dominated by excitonic transitions, two electric field regimes are possible, the so called high field regime when the applied electric fields are of the order or bigger than the exciton ionization field and the low field regime, when they are much smaller. The field regime in which one works is dictated by the ionization field of the exciton as the experimental values are confined to fields smaller than a few times 10^5 V cm^{-1}.

Patterns of behaviour characteristic of each field regime seem to have emerged from the presently available data, and for this reason EM results corresponding to both regimes will be introduced separately in the next sections.

7. Electromodulation Spectroscopy of Excitons at the Fundamental Absorption Edge in the High Field Regime

EM results corresponding to experiments performed on the exciton at the fundamental absorption edge of PbI_2, which can be considered representative of the high field regime, will first be discussed and used to build up a general model. The validity of this model will be checked by comparison of its predictions with other available results.

7.1. The exciton at the fundamental absorption edge of PbI_2

PbI_2 is one of the layer materials which has received a good deal of attention in recent years [31 to 41]. A series of lines of excitonic nature have been detected in its optical spectra at the region of the fundamental absorption edge. Originally, Nikitine et al. [31, 32] reported up to four lines; later they were verified from reflectivity measurements [39]. In order to explain them in terms of a single Wannier exciton series one has to account for the anomaly of the $n = 1$ state which appears at energies much higher than expected (discrepancies of 82 and 73 meV for the 2H and 4H polytypes respectively).

Baldini and Franchi [40] gave an interpretation of the observed spectrum in terms of two overlapping series with identical binding energies of 55 meV and separated by 24 meV. This interpretation has been questioned by Harbeke and Tosatti [42] who decided that the lines belonged to one single excitonic series, explaining the anomaly of the $n = 1$ line by a repulsive central cell correction due to the cationic nature of the exciton. However EM experiments on this exciton [43] seem to have proved that these estimates of the binding energy are too high to account for the behaviour of the EM signals.

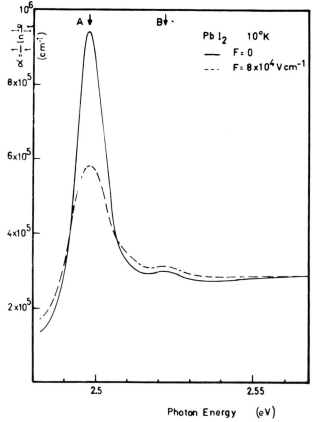

Fig. 20. Absorption spectrum of the exciton at the fundamental absorption edge of PbI$_2$ with and without an applied electric field.

In Figure 20 the absorption spectrum of this exciton is shown with and without an applied electric field. One can see that the exciton ground state, marked A in Figure 20, undergoes a large change (broadening and quenching) with its oscillator strength becoming redistributed over a relatively large energy region in the spectrum when the electric field is turned on, while the oscillator strength of the peak marked B, sometimes considered the $n = 2$ state of the same series, hardly changes at all. As the ionization field of the discrete states of a Wannier like exciton series decreases as the fourth power of the principal quantum number n, one would have expected B to have been totally erased by an electric field that has such a strong effect on peak A if they belonged to the same series. Nevertheless because of the much lower values of the absorption coefficient for the B transition, it is difficult to measure accurately the quenching of peak B by the electric field for the crystal thicknesses (~500 Å) required to measure the effect on peak A. Recent experiments [44] performed on much thicker samples (~8000 Å) seem to indicate that the field dependence of the B peak is close to that of the A peak, and no measurable dependence of its energy position with the

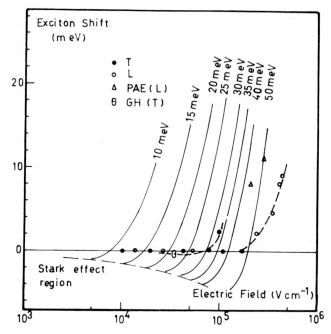

Fig. 21. Comparison of the measured shifts of the ground state of the exciton at the fundamental absorption edge of PbI$_2$ with the shifts predicted by isotropic theories for several assumptions of the exciton Rydberg.

applied electric field can be detected, which again throws some doubt about assigning the B peak to the $n=2$ state of a single excitonic series. More experimental evidence is obviously required to clarify this point.

Nevertheless it is possible to obtain a good estimate of the binding energy of the A exciton peak by studying its position as a function of the applied electric field. This is shown in Figure 21 for applied electric fields both parallel and normal to the c-axis. The points marked L and T correspond to the longitudinal and transverse geometries respectively, the point marked G—H corresponds to the shift found in ref. 39 from ER measurements for $F \perp c$ and obtained after a Kramers-Kronig analysis of the data and a careful fitting of Gaussian and Lorentzian functions to the exciton spectrum. The triangles marked PAE correspond to the shift to higher energies reported in ref. 45. The solid curves are derived from theoretical calculations [5]. One can see that the agreement between theory and experiment is quite poor. Several discrepancies are evident: at low fields there is a much smaller Stark effect than any of the curves predict and at high fields the shift to higher energies proceeds more slowly than predicted theoretically, although one notes that the experimental behaviour parallels the theoretical curve for a binding energy of around 10 meV. However this disagreement between theory and experiment is not too surprising as the theoretical curves are based on Elliot's [7] formalism which assumes an isotropic medium and consequently does not account for the anisotropy of the PbI$_2$ lattice. There are three features in Figure 21 that have to be explained (a) the near absence of the Stark effect; (b) the displacement of the exciton position towards high energies for high fields; (c) the different displacement obtained for the exciton position for fields perpendicular and parallel to the c-axis. As discussed in Section 3 the inclusion of the anisotropy leads to a reduction of the Stark effect as well as to a dependence on the orientation of the electric field [22].

Two ways of extrapolating the anisotropy into the region of high electric fields are suggested in refs. 22 and 23. Both of them allow a comparison between the experimental anisotropic curves and the isotropic theory of ref. 5, by resorting to a kind of 'isotropic equivalent parameters'. One approach relies on the mean field theory due to Pollmann [46] according to which the shift of the position of the exciton is given by:

$$\Delta E_g = (c(\gamma))^2 B \left[(C(\gamma))^{-3} \sqrt{F_\perp^2 + \frac{\mu_\perp}{\mu_\parallel} F_\parallel^2} \right]. \quad (29)$$

In which the function B is the field dependence in the isotropic case [5] but for the corrected field $c(\gamma)^{-3}\sqrt{F_\perp^2 + (\mu_\perp/\mu_\parallel)F_\parallel^2}$ which acts as the isotropic equivalent one, and where γ is the anisotropy parameter

$$\gamma = \frac{\mu_\perp \varepsilon_\perp}{\mu_\parallel \varepsilon_\parallel}$$

and $C(\gamma)$ is given by:

$$C(\gamma) = \frac{1}{\sqrt{1-\gamma}} \sin^{-1} \sqrt{1-\gamma}; \quad 0 < j < 1. \quad (30)$$

The other approach based on the variational method [22] predicts that the shifts of the exciton for applied electric fields perpendicular ($\Delta E_g^\perp$) and parallel ($\Delta E_g^\parallel$) to the c-axis are given by:

$$\frac{\Delta E_g^\perp \left(|F_\perp| \frac{E_g}{a_0} \right)}{|E_g|} = B(|F_\perp|) \quad (31)$$

and

$$\Delta E_g^\parallel \left(|F_\parallel| \cdot \sqrt{\frac{\varepsilon_\parallel}{\varepsilon_\perp} \frac{|E_g|}{a_0}} \right) + \Delta E_g^\perp \left(|F_\perp| \frac{|E_g|}{a_0} \right) = zB(|F|) \quad (32)$$

respectively. Where E_g and E_0 are given in Figure 22 as a function of the anisotropy parameter γ, and the B function has the same meaning as that in formula 29.

The black circles in Figure 23 correspond to the shift of the A exciton versus the applied electric field for $F\|c$ and $F\perp c$, while the squares correspond to the

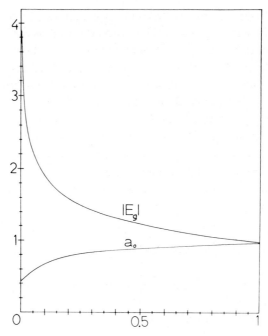

Fig. 22. Dependence on the anisotropy parameter of the E_g and a_0 parameters described in the text.

isotropic equivalent field dependence obtained after adjusting the experimental curves according to formula 29 or formulae 31 and 32. The full line is obtained from the isotropic theory of Blossey [21] which agreed very well with the corrected experimental results although there is still the disagreement in the region of the Stark effect, which is near absent in the experimental curves. Nevertheless the agreement is very good for higher fields and considering that

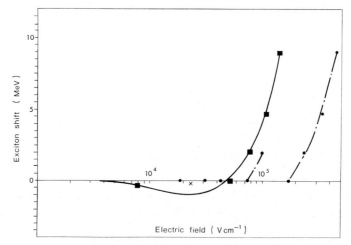

Fig. 23. A theoretical fit including anisotropy to the experimentally measured shifts induced by the applied electric fields on the ground state of the exciton at the fundamental absorption edge of PbI_2.

small values of the Stark effect are predicted by the theoretical curve (maximum of 1 meV), on the whole the agreement is very satisfactory.

A value of 16 meV for the exciton Rydberg and of 0.56 for the anisotropy parameter have been used in order to obtain this fitting, for the details of which the reader is referred to ref. 22 and 23. This value of 16 meV obtained for the exciton Rydberg provides a satisfactory explanation for the temperature broadening of the $n=1$ exciton line found experimentally, which the higher values proposed are unable to give. Roughly the effect of temperature is to increase the broadening of the exciton line as kT, at least above, say, liquid nitrogen temperatures, where temperature becomes the main broadening mechanism other than externally induced ones, like applied electric fields or induced disorder (ion bombardment for instance). For a Rydberg of 20 meV the broadening parameter $\Gamma = kT$ at room temperature should give Γ/R close to unity and, as shown in Figure 2 in Section 2, one should expect to see just a faint peak for the $n=1$ state as found experimentally. For higher values of R (say 127 meV), then $\Gamma/R \sim 0.2$, and a well defined exciton peak should be seen, in disagreement with the experimental evidence.

Typical EA spectrum of this exciton at 80 K and for $F \perp c$-axis is shown in Figure 24. It is worth pointing out that for small fields the redistribution of the oscillator strength when the field is applied, yields a signal with a relatively larger response at the high energy side of the field free position of $n=1$. This effect is due (and it will become more evident in Section 12, when results on the low field regime are presented) to the fact that as the ionization field for higher order states is much smaller than for $n=1$, they are more affected than the ground state for small fields. In other words when one increases the electric field from zero the effects are first felt in the continuum and quasicontinuum region. As the field increases these states become totally ionized and the response from $n=1$ start taking over. When fields of the order of the ionization field for $n=1$ are applied, the type of response obtained is one large negative peak at the energy position of $n=1$, with positive satellites at each side and oscillations extending well into the region of the continuum.

As it will be shown in Section 12 with more clarity, when the binding energy of the exciton is large enough so that Γ/R is small, one can detect definite structure in the EA signal arising from higher order discrete states. One can say that the type of signal shown in Figure 24 is typical of the high field regime [47], this is the EM type of spectra one should expect to obtain for applied electric fields of the order of or smaller than F_I.

The way in which the several features in the EA spectrum behave under the applied electric fields, for the applied field $F \perp c$ is illustrated in Figures 25 and 26. Figure 25 shows how the amplitudes of the EA peaks, obtained for $F \perp c$, as a function of the strength of the applied fields. It can be seen in Figure 25 that for fields smaller than 8×10^4 V cm^{-1} a power law is followed. For larger fields a trend towards saturation appears.

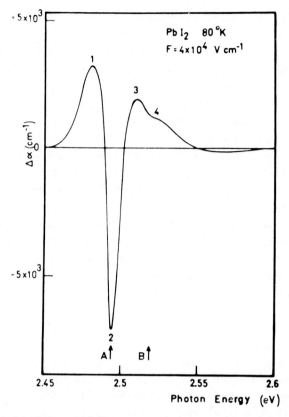

Fig. 24. A typical electroabsorption spectrum at 80 K and an applied electric field of 8×10^4 V cm^{-1} yielded by the exciton at the fundamental absorption edge of PbI$_2$.

The upper part of Figure 26 shows how the position of the EA peaks change with electric field perpendicular to the c-axis. For fields smaller than 8×10^4 V cm^{-1} the positions remain practically unchanged, but for larger fields, peak 1 starts shifting fairly rapidly to lower energies, peak 3 to higher energies, while peak 2 remains essentially unchanged in position.

In the lower part of Figure 26 the separation in energy of the zero line crossing points, shown as ΔE_1 and ΔE_2, are plotted versus the applied field. ΔE_1 does not change for fields smaller than 8×10^4 V cm^{-1}, but for bigger fields it starts to change rapidly. ΔE_2 changes quite slowly for fields smaller than 8×10^4 V cm^{-1} but it starts moving fairly rapidly above this value. The fact that the width E_2 is partly determined by a contribution from the effects of the electric fields on the higher order states is probably responsible for the slow variation shown for fields smaller than 8×10^4 V cm^{-1}.

Observe that the critical value of the electric field $F_c = 8 \times 10^4$ V cm^{-1} for which these effects are noticeable coincides quite closely for the value of the electric field that makes the exciton peak move towards higher energies. In other words,

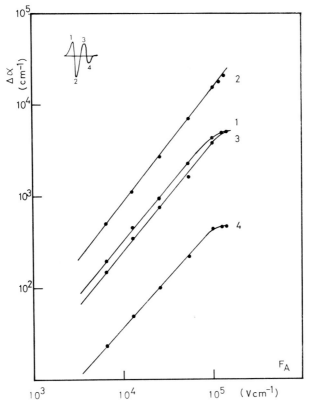

Fig. 25. Electric field dependence of the amplitude of the peaks of the electroabsorption spectrum of the exciton at the fundamental edge of PbI_2 at liquid nitrogen temperature and for $F \perp c$-axis.

for a field such that the exciton gathers enough energy for the electron to disengage itself from the Coulombic well.

For fields applied parallel to the c-axis a similar behaviour is obtained except that the trend towards saturation does not appear until fields of the order of 10^5 V cm^{-1} are reached.

Thus it seems possible that from this type of measurement one can obtain accurate values of F_I for several preferential directions within a layer lattice which in turn can be used to work out accurate values of the excitonic parameters and of their anisotropies.

7.2. A MODEL FOR ELECTRIC FIELD EFFECTS ON BIAXIAL EXCITONS IN THE HIGH FIELD REGIME

The main conclusions obtained from the EA experiments on the exciton at the fundamental absorption edge of PbI_2 can be summarized as follows:

(a) If the anisotropy parameter of the lattice ($\gamma = (\varepsilon_\perp \mu_\perp / \varepsilon_\parallel \mu_\parallel)$) is appreciably different from unity, the Stark effect will be greatly reduced and consequently

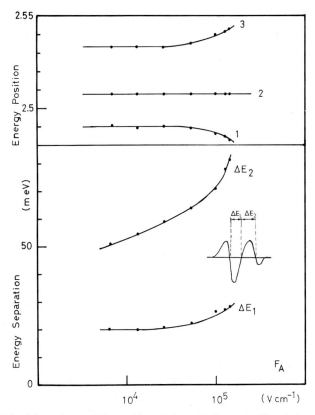

Fig. 26. Electric field dependence of the position of the peaks (upper half) and the energy separation of the zero line ironing points (lower half) of the electroabsorption spectrum of the exciton at the fundamental absorption edge of PbI$_2$ at liquid nitrogen temperature and for $F \perp c$-axis.

difficult to measure unless the exciton Rydberg is large. The most noticeable effect will be that for larger applied fields the exciton ground state energy position will shift to higher energies quite rapidly, which in principle one can fit with the help of the present available theories [22] and obtain an estimate of the exciton parameters involved.

(b) The amplitudes of the EA peaks increase with the applied electric fields following a power law up to a certain critical field F_c close in value to F_I; when larger fields are applied the amplitudes tend to saturate.

(c) The energy position of the EA peaks remains the same for fields smaller than F_c; for higher fields the two main satellite peaks move to lower and higher energies respectively.

(d) The zero-line crossing point energy separation ΔE_1 is independent of the applied field if this is smaller than F_c, for bigger fields it increases rapidly. ΔE_2 does change as a function of the applied field even if this is smaller than F_c but when the applied field reaches the critical value the rate of change of ΔE_2 suddenly increases.

(e) All these effects described in (a), (b), (c) and (d) will show the anisotropy of the lattice when the applied fields have different orientations.

In order to check the general validity of this model, some EM results obtained on excitons in layer materials are presented in the next section and will be discussed and compared with the main conclusions of this model.

8. The Excitons at the Fundamental Absorption Edge of GaS, GaSe and GaTe

Of the group of the $A^{III}B^{VI}$ layer compounds, GaS, GaSe and GaTe are the ones which have probably undergone the most extensive study with the use of modulation techniques.

GaS and GaSe crystallize in the trigonal prism coordination and can so be compared from the crystal structure point of view with MoS_2. These crystals consist of rather loose stacks of mainly covalently bonded layers, each of which contains four monoatomic sheets. The crystal structure of GaS and GaSe differ only in the way the four-fold atomic layers are stacked. The crystal lattice of GaTe has a rather more complicated monoclinic structure which is responsible for the fact that no satisfactory band structure calculation is as yet available.

As in most layer materials, the bonding between the layers is of the Van der Waal type and it is believed that their optical properties are largely governed by the individual fourfold layers.

The selection rules for the optical transitions of these materials are such that if the polarization vector of the light is in the plane of the layers the transitions governing the fundamental absorption edge give rise to only a relatively small absorption and it is therefore possible to carry out modulated absorption measurements with relative ease well beyond the fundamental edge. The easy cleavage of these materials leads to good surfaces for reflectivity measurements. A large portion of the modulation work carried out on these materials is concerned with the excitonic transitions at the fundamental absorption edge which all these materials show.

The first modulation experiments performed on GaSe concerned EA [48] and ER [49]. The EA experiments used a transverse geometry while a longitudinal geometry was employed for the ER experiments (modulation was achieved by the interface barrier potential of a GaSe—SnO_2 hetero-junction).

Figure 27 shows the EA of the exciton at the fundamental absorption edge of GaSe [48] at 395, 345, 290 and 80 degrees Kelvin respectively. The applied electric field (F) had a constant (F_0) and an alternating component (F_1):

$$F = F_0(1 + a \text{ cso } wt); \qquad a = \frac{F_1}{F_0}. \tag{33}$$

The data shown in Figure 27 was obtained with $F_0 = 1.6 \times 10^4 \text{ V cm}^{-1}$ and $a = 0.7$. The signal was detected at the frequency of the applied electric field.

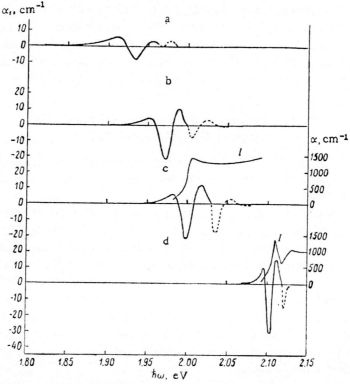

Fig. 27. Amplitude of the first harmonic, α_1, as a function of the energy of the light quanta. (a) $T = 395$ K, $F_0 = 1.6 \cdot 10^4$ V cm^{-1}; (b) 345 K, $1.6 \cdot 10^4$ V cm^{-1}; (c) 290 K, $1.6 \cdot 10^4$ V cm^{-1}; (d) 80 K, $1.4 \cdot 10^4$ V cm^{-1}; in all cases $a = 0.7$. The scale of the curves shown as dashed lines is increased by a factor of 10. I: The absorption spectrum.

After considering the dependence of the EA spectra on temperature and applied electric fields, the authors had to conclude that the F–K effect failed to explain the results and they had to invoke exciton ionization by the electric field to account for the behaviour of the EA spectrum.

Exciton effects are responsible for the fact that the amplitudes of the EA peaks fall off more rapidly for increasing temperatures than the F–K effect predicts. In addition the rapid broadening with temperature of the EA signal disagrees with the predictions of the F–K effect.

The type of signal shown in Figure 27 is clearly similar to those yielded by PbI_2 and discussed in Section 7.1, the large negative peak at the energy position of the exciton ground state with the two satellites at each side corresponds to the shape yielded by an exciton when it is quenched and shifted to higher energies by an electric field in a manner typical of the high field regime. The field dependence of the amplitudes of the EA peaks was found to follow a power law with the exponent close to unity. Again this power law is typical of the high field regime.

The authors of the work under discussion tried to obtain an estimate of the binding energy of the exciton involved, by attributing the second negative peak of the EA spectrum to the electric field induced quenching of the $n = 2$ state. Fitting a hydrogenic series they estimate an exciton Rydberg of 70 ± 10 meV at liquid nitrogen temperatures. Nevertheless it is more likely that the second peak is associated with the expected oscillations which appear in the region of the continuum due to the redistribution of the oscillator strength of the exciton under the applied electric field. In order to achieve an accurate estimate of the exciton parameter by means of EM experiments one should attempt to measure the field dependence of the energy position of the exciton ground state and analyse the results including the effect of the anisotropy along similar lines to those described in Section 7.1 for PbI_2.

Recent estimates for the value of the exciton Rydberg in GaSe [50] give a value for the binding energy of $n = 1$ of $R = 24.8$ meV and an effective Bohr radius of around 32 Å which would just yield the sort of line shapes shown in Figure 29.

Figure 28 shows the electric field induced change in the imaginary part of the dielectric constant E_z of the exciton at the fundamental absorption edge of GaSe as obtained by Suzuki et al. [49] by a K–K analysis of their ER data which is also shown in Figure 28. The line shape shown in Figure 28 differs quite considerably from the one shown in Figure 27.

Figure 28 has three negative peaks instead of the two shown in Figure 29. The authors attributed the peak 'a' to the existence of an impurity level created during the process of sample preparation and associated 'b' and 'c' with the $n = 1$ and $n = 2$ discrete states respectively, obtaining a binding energy of around 23 meV. Although this is a value close to other independent estimates of the binding

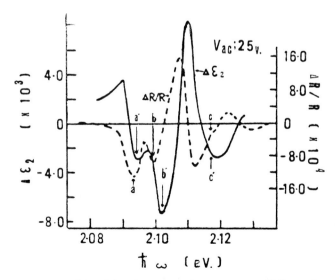

Fig. 28. An imaginary part of the dielectric constant $\Delta\varepsilon_2$ spectrum at 90 K transformed from $\Delta R/R$ through Kramers-Kronig relation (solid line) and electroreflectance spectrum $\Delta R/R$ measured at 90 K (dashed line).

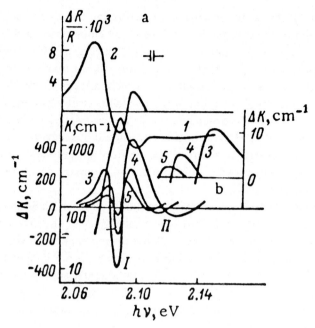

Fig. 29. Absorption (1), electroreflection (2) and electroabsorption (3–5) spectra of a GaSe single crystal. E (V cm^{-1}): (2) 3×10^4; (3) 3.2×10^4; (4) 2×10^4; (5) 9.5×10^3.

energy it is more likely that peak 'c' is associated with the oscillations appearing in the region of the continuum. If peak 'a' was not present in the spectrum shown in Figure 28 the line shape would be the same as that in Figure 27.

Figure 29 shows the EA spectra of the exciton at the fundamental absorption edge of GaSe at 77 K for $F \| c$-axis and the ER spectrum for $F \perp c$-axis [51]. The EA line shapes are very similar to those in Figure 7, but two effects can be observed in Figure 29: (a) For increasing values of the applied electric field the energy positions of the EA peaks change. The first positive peak moves to lower energies while the second one and the oscillations after it move to higher energies. The first negative peak remains essentially at the energy position of the exciton ground state. (b) The period of the oscillations of the EA signal gets bigger for increasing fields. The effect is noticeable, indicating that the ionization field has been reached. From these two effects one can estimate that F_I must be around 2×10^4 V cm^{-1}. Assuming a binding energy for the ground state of 24.8 meV [50] the exciton radius along the c-axis should be 124 Å. This seems a rather large value for the Bohr radius and it might be that the values of the applied field were over-estimated. As this result seems to suggest that this exciton is more extended perpendicular to the layers than parallel to them, (a rather unusual situation), an anisotropy parameter bigger than unity is implied.

Figure 30 and Figure 31 show the EA and ER spectra of GaS (77 K [51], $F \| c$ and $F \perp c$ for the EA and ER respectively) and GaTe (300 K [52], $F \perp c$)

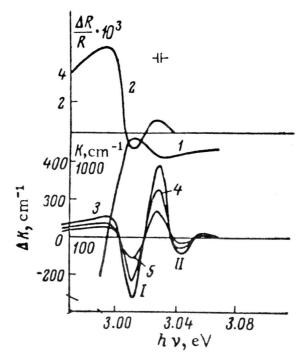

Fig. 30. Absorption (1) electroreflection (2) and electroabsorption (3–5) spectra of a GaS single crystal. E (V cm^{-1}): (2) 4×10^4; (3) 6×10^4; (4) 4×10^4; (5) 2.3×10^4.

respectively, in the region of the exciton at the fundamental absorption edge. A comparison of the EA spectrum of GaS (Figure 30) with the one for GaSe (Figure 29) reveals the similarity between the two spectra but one can appreciate subtle differences: (a) the EA peaks in GaS are broader than in GaSe, but they do not broaden with increasing electric fields, (b) The position of the EA peaks remains at the same photon energies for increasing applied fields.

As both experiments were performed under the same experimental conditions ($F \perp c$ and at 77 K) according to the model described in Section 9, one should take these results as an indication that the ionization field of the exciton at the fundamental absorption edge is higher for GaS than for GaSe – in other words the binding energy is probably bigger.

The EA data on the exciton at the fundamental absorption of GaTe shown in Figure 31 was obtained at 300 K with $F \perp c$. The applied electric field had a d.c. and an a.c. component: $F = F_0 + F_1 \cos wt$ for the spectra shown in Figure 31. $F_0 = 1.1 \times 10^4$ V cm^{-1} and constant for all the shown curves. With F_1 varying as follows:

(1) 3.3×10^3 V cm^{-1}
(2) 5.5×10^3 V cm^{-1}
(3) 7.7×10^3 V cm^{-1} and
(4) 9.9×10^3 V cm^{-1}.

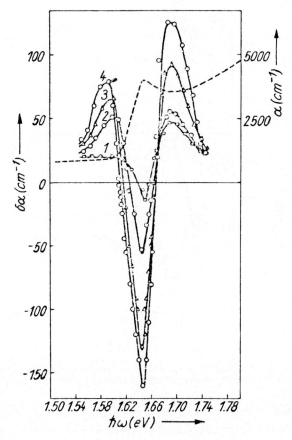

Fig. 31. Absorption coefficient α as a function of photon energy $\hbar\omega$ in the absence of an electric field (dashed line), and $\delta\omega$ as a function of $\hbar\omega$ in the exciton absorption region for $F_0 = 1.1 \times 10^4$ V cm^{-1} and different fields F_1: (1) 3.3×10^3, (2) 5.5×10^3, (3) 7.7×10^3 and (4) 9.9×10^3 V cm^{-1}.

The fact that the widths and positions of the EA peaks vary with increasing fields is by itself an indication of smaller F_I than in the case of GaS according to the model described in Section 9.

In Figure 32 the dependence on EA peaks for a constant $F_1 = 6 \times 10^3$ V cm^{-1} and varying F_0 is shown [52]. According to the EM model described in Section 5 the change $\Delta\alpha$ can be considered to be under these experimental conditions:

$$\Delta\alpha = \alpha(F_0 + F_1) - \alpha(F_0 - F_1). \qquad (34)$$

The fact that a saturation level is reached and under the assumption that no internal fields were present (which was not checked by the authors), means, according to our model (Section 5) that $F_0 + F_1 = F_I$. As this happens for $F_0 = 9 \times 10^3$ V cm^{-1} and $F_1 = 6 \times 10^3$ V cm^{-1} [53] then $F_I = 1.5 \times 10^4$ V cm^{-1}.

Several estimates of the binding energy of this exciton have been made [53] and although discrepancies are found, the value $R = 14$ meV would mean, for this F_I,

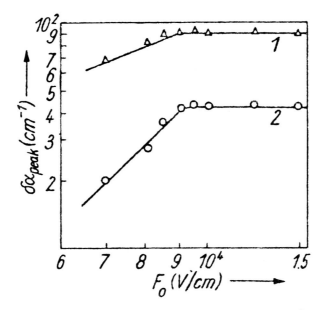

Fig. 32. Dependence of $\delta\alpha_{peak}$ on the constant electrical field F_0. $F_1 = 6 \times 10^3$ V cm^{-1}. (1) second positive peak, (2) negative minimum.

an exciton effective Bohr radius of about 92 Å, which seems a reasonable enough value. However in this analysis the effects of anisotropy have been neglected and these would certainly alter the picture obtained as in the case of PbI$_2$.

It is obvious that in order to carry out a comprehensive analysis along these lines, careful measurements for the several possible orientations of the electric field should be undertaken, with modulation achieved from the zero internal field position and at temperatures as low as possible in order to minimize unnecessary interferences. Nevertheless in the light of the presently available EM data on the excitons at the fundamental absorption edges of GaS, GaSe and GaTe, it seems that they do behave according to the model proposed in Section 9 and at the same time they verify the fact that the exciton at the fundamental absorption edge of PbI$_2$ has a Rydberg comparable to the ones for GaS, GaSe and GaTe, with all the experiments falling in the category of EM on excitons in the high field regime.

9. Electromodulation Spectroscopy of Excitons at the Fundamental Absorption Edge in the Low Field Regime

In an analogous manner to PbI$_2$, 2H—MoS$_2$ is chosen as a representative layer material with an exciton at the fundamental absorption edge which exhibits a behaviour that one can consider typical of the low field regime (F applied $\ll F_I$) and the EM results on this exciton will be used to build up a model which explains at least qualitatively the experimental results on other transition metal dichalcogenides. Besides, 2H—MoS$_2$ exhibits some of the effects discussed in Section 5, and it provides a good example on how to extricate the relevant information from the EM data.

The A exciton at the fundamental absorption edge of 2H—MoS$_2$ has a fundamental absorption edge dominated by a pair of strong exciton transitions, named A and B, which are assumed to originate from a spin-orbit split valence band.

Along with the other group VI transition metal dicholcogenides, it is thought to be a narrow band semiconductor [54] (with an electric energy gap of about 0.25 eV) and it is known to exist as both n-type and p-type, depending on the history of each particular crystal. Nevertheless the value of the electric gap should be taken with some reservation, as reliable electrical measurements are difficult to achieve due to the somewhat delicate layer structure and the considerable anisotropy.

The nature of the uppermost filled band and the lowest empty band forbids strong optical transitions between them, and the fundamental absorption edge occurs at a higher energy than the electrical gap (at around 1.9 eV as shown in Figure 33). As the A and B exciton series of 2H—MoS$_2$ do not appreciably overlap and the A exciton has a binding energy of the order of 55 meV with

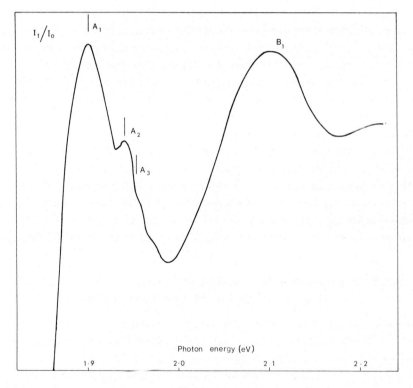

Fig. 33. The absorption coefficient of 2H—MoS$_2$ in the region of the fundamental absorption edge at 10 K.

$F_I \sim 2 \times 10^5$ V cm^{-1}, 2H—MoS$_2$ is a good candidate for the study of the effects of electric fields in the low field regime.

The EA results reported were performed with the longitudinal method [30, 55] and for the reasons suggested in Section 5 the quoted values of the applied electric field are likely to be overestimated.

9.1. Electromodulation experiments on the A exciton at the fundamental absorption edge of 2H—MoS$_2$

The photon energy dependence of the absorption coefficient obtained in the energy region of the fundamental absorption edge for a crystal of MoS$_2$ at 10 K is shown in Figure 33 [30]. This spectrum shows clearly the $n = 2$ and $n = 3$ discrete states for the A series. The beginning of the continuum can be obtained by fitting the hydrogenic formula:

$$E_n = E_\infty - \frac{\mu e^4}{2h^2 \varepsilon_x^2 n^2} \tag{35}$$

(μ is the reduced effective mass of the electron-hole pair, ε_x is a dielectric constant) to the series after taking crystal thickness or size effects into account [57]. No higher order discrete states could be detected for the B series, and the half width of B ($n = 1$) is much larger than that of A.

All of the electroabsorption (EA) spectra to follow were obtained at about 80 K [30]. The full line of Figure 34 shows the EA spectrum obtained at the same frequency (S_f) as the applied voltage, which was sinusoidal about zero, for a 500 Å thick crystal. The frequency f was 300 Hz and the calculated field F_A equalled 2.5×10^5 V cm^{-1}. The broken line corresponds to the signal at the first harmonic (S_{2f}) obtained under the same conditions. The energy position of $n = 1$ and $n = 2$ indicated by arrows, were found by simultaneous measurements of transmission.

The various peaks in the EA spectrum are numbered in order of increasing energy. The shapes of S_f and S_{2f} are very similar, differing only slightly in the location of peaks 1, 2, 7. Peaks 0, 1, 2 are clearly related to the effect arising from the $n = 1$ state, although some contributions to peak 2 might come from higher order states. As $\Delta \alpha$ is shown here, one might say that the signal found around $n = 1$ corresponds to *a shift to higher energies of the exciton ground state* and, in the case of S_{2f}, a similar shift accompanied by a *slight enhancement* in α at the peak of $n = 1$. (It will be shown later that this would be an incorrect conclusion). The oscillation that gives rise to peaks 3, 4, 5 and which appears in the energy regions of $n = 2$ is clearly associated with this exciton discrete state, but the large oscillation 5, 6, 7 appear in the region of the exciton continuum (i.e. the 'band edge', if one uses a term which is strictly meaningful only for one-electron models). We see that the biggest response comes from the 'band edge'. The ratio S_{2f}/S_f in this particular case is smaller than one, (except for peaks 2 and 4) but its

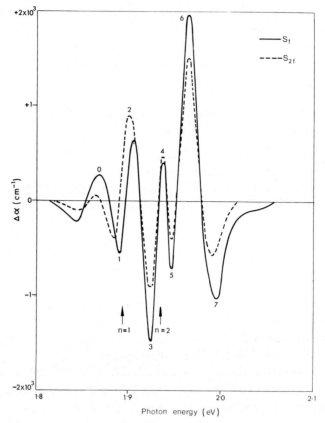

Fig. 34. Electroabsorption spectra of the A exciton of 2H—MoS$_2$ obtained at the frequency of the applied sinusoidal electric field (full line) and at the first harmonic (broken line). ($F_A = 2.5 \times 10^5$ V cm^{-1}, 80 K).

value proved to be different from sample to sample, S_{2f} being sometimes bigger, sometimes very much smaller than S_f.

As the amplitude of the sinusoidal field increases, peaks 0 and 1 diminish in size, 4 and 5 mix together and strongly overlap. The EA spectrum found for an applied field with a peak to peak value of $F_A = 10^6$ V cm^{-1} is shown in Figure 35. A surprising result is that the signal around $n = 1$ has disappeared almost completely, leaving only an EA response at higher energies. Theoretical predictions (see Section 3) indicate that as the electric field increases, the EA response from $n = 1$ should become stronger.

A rather different type of signal was obtained when modulation was achieved with the use of an alternating square wave voltage describing positive excursions from zero to a value $F_A = 1.6 \times 10^5$ V cm^{-1} (see Figure 36(a)). In this case the largest signal in the spectrum appears around $n = 1$, and corresponds to the shape expected for a shift of the exciton ground state to lower energies accompanied by some broadening. A small oscillation (numbered 3', 4') follows in the energy

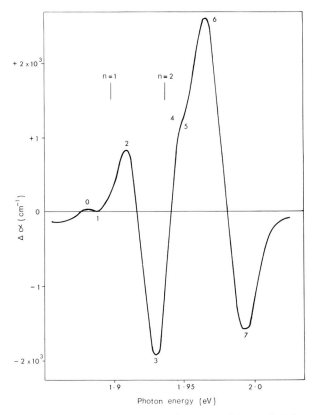

Fig. 35. Electroabsorption spectrum obtained at the frequency of the applied sinusoidal electric field. ($F_A = 10^6$ V cm^{-1}, 80 K).

region of $n = 2$ and a response, completely different from the one found for sine wave modulation, appears at higher energies.

On applying a d.c. biasing field (F_B) so as to make the electric field oscillate during one half cycle to negative values the signal disappeared for a certain value of F_B in between 0 and $-F_A$ (in fact $\sim -0.8 F_A$) before reappearing and developing to the shape illustrated by Figure 36(b). For $F_B = -F_A$ a change of sign in the region of $n = 1$ has been achieved without changing the phase of F_A. The signal obtained in this case looks very much like the one found for sine wave modulation. Again, in the present case, the signal around $n = 1$ (peaks 0, 1, 2) has the shape corresponding to a shifting of the exciton ground state towards higher energies. An inversion of sign is also seen in the energy region of $n = 2$. However, the main difference between the full and dotted lines, leaving aside the inversions of sign, is the appearance of the large oscillations 6, 7 in the energy region of the exciton continuum.

The way in which the sizes of the several peaks, obtained for an applied square wave electric field $F_A = 1.6 \times 10^5$ V cm^{-1}, as a function of the applied d.c. bias (F_B

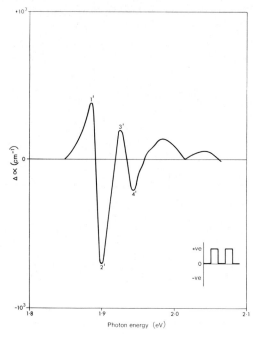

Fig. 36. Electroabsorption spectra of the A exciton of 2H—MoS$_2$ for a square wave applied electric field ($F_A = 1.6 \times 10^5$ V cm^{-1}, 80 K) (a) For the applied field describing positive excursions (b) For the applied field describing negative excursions.

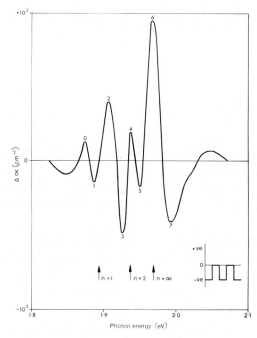

Fig. 36b.

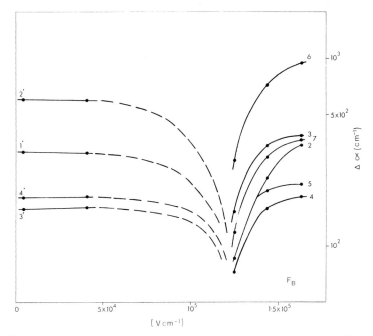

Fig. 37. Amplitudes of the peaks of the electroabsorption spectra of the A exciton of 2H—MoS$_2$ for a fixed $F_A = 1.6 \times 10^5$ V cm^{-1} as a function of the biasing field F_b.

between 0 and $-F_A$) is shown in Figure 37. As F_B increases, the size of the peaks 1', 2', 3', 4', decrease until no signal is detected at $F_B = 1.2 \times 10^5$ V cm^{-1}. For larger F_B, peaks 2 → 7 appear and increase in size as shown. The critical value of F_B for which $S_f = 0$, varied from sample to sample, and, occasionally, inversion could not be obtained.

For the two conditions of biasing shown in Figure 36 ($F_B = 0$ and $F_B = -F_A$), the amplitude of the peaks was measured as a function of the magnitude of the alternating field F_A. The results are shown in Figure 38. For negative excursions of the applied field, ($F_B = -F_A$) peaks 2 → 7 increase with F_A following a power law up to a value of $F_A = 2 \times 10^5$ V cm^{-1}, above which the size of the peaks decrease. The latter effect is very drastic for all the peaks except 2; the rate of growth of peak 1 was non-monotonic and could not be followed properly. For high values of F_A there is a tendency for the signal around $n = 1$ to overcome in size the remaining ones (as shown by representative-peak 2, in Figure 38). For positive excursions of the applied field ($F_B = 0$), 1' and 2' increase in size following a power law with field and they do not show any tendency towards saturation. The variations in the amplitudes of peaks 3' and 4' with F_A are not shown in this case, as these have a tendency to suffer overlap effects at high fields.

These experimental results can be interpreted with the help of the model described in Section 5.

We shall first attempt to interpret the results obtained using square wave

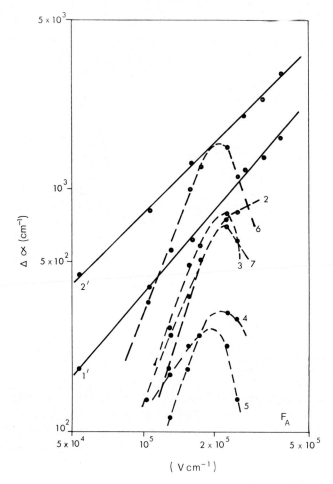

Fig. 38. Amplitudes of the peaks of the electroabsorption spectra of the A exciton of 2H—MoS$_2$ as a function of the applied electric field F_A, for $F_b=0$ (full lines) and $F_b=-F_A$ (dashed lines).

modulation i.e. those shown in Figures 36(a) and 36(b). The change in sign of the observed signal in the region of $n=1$ and $n=2$ shown in Figure 36 for the two waveforms shown in the insert suggests that modulation is occurring from a point close (relative to the applied alternating field of magnitude 1.6×10^5 V cm^{-1}), to the zero field condition. However the fact that the signal goes to zero for a value of $F_B \neq F_A/2$ (Figure 39) indicates the existence of a small internal field. In this particular case the value of F_B that makes the signal zero is bigger than $F_A/2$ and it is inferred that there is a positive internal field F_i given by:

$$F_i = F_B - F_A/2 = 1.2 \times 10^5 - 0.8 \times 10^5 = 4 \times 10^4 \text{ V cm}^{-1}.$$

The above estimate of the internal field is based on the assumption of negligible screening by surface states and is therefore likely to be an overestimate. As we

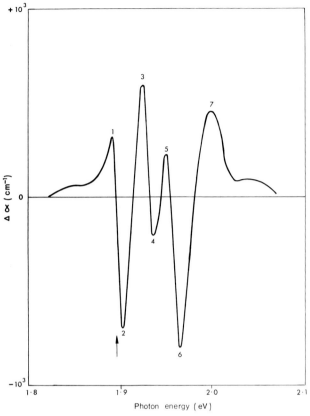

Fig. 39. Electroabsorption spectrum of the A exciton of 2H—MoS$_2$ at 80 K showing $\Delta\alpha = \alpha(2.5\times 10^5 \text{ V cm}^{-1}) - \alpha(0)$.

shall see below, it appears that the calculated fields are too high by at least a factor of 3.

If by fitting a Wannier series, we now take the exciton Rydberg to be 55 meV (see Figure 33) and the dielectric constant 6.8 [58], the effective Bohr radius using the hydrogenic formula, is ~19 Å. The ionization field for this state is thus $F_I = 2.9\times 10^5$ V cm^{-1}. Under conditions of maximum field that could be applied before breakdown, namely $F_A = 4\times 10^5$ V cm^{-1}, no decrease in the size of the signals for positive applied fields was detected (see Figure 38). For this field ($F_A/F_I \sim 1.4$), the theoretical curve of Figure 8 indicates that not only should the signal have decreased, but a change of sign and a strong shift of the exciton to higher energies should have resulted. As this was not observed we conclude that the maximum applied field corresponds to no more than $F_A/F_I \sim 0.5$ and thus $F_A < 1.5\times 10^5$ V cm^{-1}. We thus obtain our estimate that the calculated fields are too high by at least a factor of 3 and possibly much more.

The change of sign suffered by S_f around $n = 1$ and $n = 2$ when a biasing field is applied, as illustrated by Figure 36(a) and Figure 36(b) can therefore be explained, in terms of having modulation from near the flat band (zero internal field) position. Under these conditions, the two types of applied field illustrated in the insert to Figure 36 lead to effective fields in the sample which swing from close to zero to positive and negative values respectively. An effective change of phase in the effective field therefore occurs and $\Delta\alpha$ changes sign according to the definition given in Section 5.

The difference in the appearance of the large signal in the energy region of the continuum can be explained by differences in the penetration depth of the effective fields for the two polarities. It has been shown [57, 59] that for crystal thicknesses of the order of the ones used in this work, the exciton states undergo a significant shift towards higher energies as one approaches the surface of the crystal. The higher the order of the state the more important is the effect. This shift should be accompanied by an increased probability of ionization for the discrete states. Unfortunately, no quantitative calculations are available on this latter effect, but qualitative considerations indicate that, for crystals of the size used, the effect should be quite important. Of course, the critical distance from the surface of the crystal, within which a given exciton state will be ionized, depends on its spatial extension and accordingly the higher the order of the state the bigger this distance will be. If, for example, the penetration depth of the electric field was smaller than, let us say, the critical distance for $n = 3$, then one would not expect any change in α associated with any discrete state with $n > 3$.

It is therefore suggested that positive excursions of the applied voltage lead to accumulation layers, negative excursions to depletion or inversion layers. For the former case the penetration depth would be much smaller than for the latter; as a result states of large radii would not be affected and the signal in the region of the quasicontinuum would be effectively missing as observed.

A rough estimate of the parameters involved can be made in order to test this suggestion. Taking the electrical band gap to be 0.25 eV i.e. 36 kT at 80 K, the carrier concentration (electrons) to be $n_c \sim 10^{15}$ cm^{-3} (the order of magnitude obtained by electrical measurements), and the effective density of states in the conduction band to be 10^{19} cm^{-3}, the separation of the Fermi level from the conduction band edge is $(E_c - E_F)/kT \sim 10$. The bulk potential $u_B = (E_F - E_i)/kT$, where E_i is the mid-gap energy, is then 9. The effective Debye length is given by $L \sim (\varepsilon kT/2e^2 n_c)^{1/2}$ i.e. for $\varepsilon = 6.8$, $L \sim 1200$ Å.

The maximum barrier height at the surface for the case of an accumulation layer will be taken in units of kT as $v_s = 10$ i.e. when the conduction band at the surface reaches the Fermi level. The effective depth of the space charge region i.e. the penetration depth of the field is then $\sim 0.05L$ i.e. ~ 60 Å [60]. For an inversion layer the penetration depth reaches its maximum value when $v_2 = 2u_B$. In our case ($u_b = 9$) this is $\sim 3L$ i.e. ~ 3600 Å [60]. The applied fields are certainly sufficient to achieve these barrier configuration unless screening by surface states is very significant. For the case of negatively applied voltages it is possible that the maxima shown in Figure 38 could correspond to the attainment of the maximum penetration depth.

The sample thickness is ~ 500 Å. Thus the accumulation layer calculated above extends only a short distance into the sample; on the other hand the inversion layer completely penetrates the sample, leading to an effectively uniform field. Evans and Young [58] give the ground state exciton radius to be 24.5 Å perpendicular to the layers and 38.5 Å parallel to the layers. For the thin crystals the average radius is probably closer to 20 Å. Taking a value of $25n^2$ Å for the radius of the nth state, it is clear that when modulation occurs from flat band to accumulation layers it is extremely unlikely that high order exciton states will be significantly affected; because of their large radii they will not be found in the region near the surface where the electric field is mainly active.

This is the interpretation proposed for the effective absence of a band edge signal in Figure 36(a). It is also the reason why the dotted curve is taken to represent the best experimental conditions (the most uniform field) for comparison with theory.

The position of $n = \infty$ indicated by an arrow in Figure 36(b) falling at about the energy of the large peak 6 is calculated after correction to the hydrogen-like formula for the thickness of the crystal used. The energy separation between $n = 1$ and $n = \infty$ is about 74 meV which is to be compared with the value of about 55 meV expected for thick crystals which do not exhibit a size effect (it should be mentioned that even with the high degree of anisotropy existing in MoS_2, the hydrogenic relation holds well at least up to $n = 3$). It is to be emphasized, however, that the coincidence of peak 6 with the energy position of $n = \infty$, is probably accidental and in general no feature of the EA spectrum can be taken as a marker for the onset of the continuum.

Because of the definition of sign used for $\Delta\alpha$, Figure 36(b) (and also both lines of Figure 34) shows an apparent shift of the exciton ground state toward higher energies, but if the real effect of the field on $\Delta\alpha$, this is $\Delta\alpha = \alpha(F_A) - \alpha(0)$, had been plotted, then the sign of $\Delta\alpha$ would be inverted. Figure 39 shows the EA spectrum for the A exciton for an applied field describing negative excursions and $F_A = 2.5 \times 10^5$ V cm^{-1} with the sign of $\Delta\alpha$ given as $\Delta\alpha = \alpha(F_A) - \alpha(0)$. (The small value of F_i is not considered to be important). It is seen that the signal found at $n = 1$ corresponds to a shift to *lower* energies, this being the actual effect of the electric field on the exciton ground state.

The ionization field is $F_I = 2.9 \times 10^5$ V cm^{-1} and in the case given by Figure 39, $F/F_I = 0.86$ which, according to the theory would correspond to the exciton ground state remaining at the same energy position but broadened. However, as it has already been suggested it is quite likely that the values of F_A have been systematically over-estimated as the effect of surface states has not been included; this would agree with the theory. Naturally, anisotropy and size effects should be included in the theory before quantitative comparisons with experiment are attempted.

In the light of the given interpretation of the EA spectrum obtained for square wave modulation, the signals obtained with sine wave modulation (Figure 34 and Figure 35) can be better understood. In this case the signal arises mainly from the half cycle which has the large penetration depth. Similarity with Figure 36(b) becomes clear. The signal at f and at $2f$ appear for two reasons: (a) The existence of an internal field and (b) the distortion in the sine waves. Only through a careful analysis can one work out the several contributions assuming that parameters like F_i and the field penetration depth are known. But as has been shown this difficulty can be overcome with use of square wave modulation combined with biasing studies.

Another point is that using sine wave modulation, for high values of the applied field, the effect on $n = 1$ of the positive half cycles and the negative half cycle will tend to cancel each other, explaining the disappearance of the signal around $n = 1$ for high fields as shown in Figure 35.

Finally, it should be stated that each particular crystal of MoS_2, has its own characteristic EA spectrum, in the sense that F_i can change from sample to sample as well as the carrier concentration and the type of majority carrier (n or p type). Experimentally different values of F_i can be detected ranging from very small to very large values. However, in every case the behaviour encountered is always consistent with the proposed interpretation.

The origin of the internal field F_i is not clear at the moment. It seems likely that it arises from the contact potential between the MoS_2 and the metal electrodes, but it could be present on a free surface due to charge in surface states.

A different interpretation for the EM signal yielded by the A exciton in $2H$—MoS_2 has recently been proposed [61] based on electroreflectance results performed with the transverse geometry. The ER spectrum obtained by the author of ref. 61 at 77 K and for an applied field of 500 V cm^{-1} is shown in Figure 40. Based on the observations that (1) the size of the ER signal is relatively large for the field applied; (2) that the signal was detected at the same frequency as the applied electric field (although symmetric electrodes were used); (3) the size of the signal reaches a maximum at 50 K and decreases with decreasing temperature until at 10 K it has the same size as at 77 K, it was proposed that the spectrum is due to piezoelectric effects. A strong coupling of the electric field with the lattice results in a change in the local fields by a deformation and hence a polarization of the lattice which in turn affects the electronic states of the crystal. This interpretation implies a considerable degree of ionicity in $2H$—MoS_2 which, in principle, is plausible as it is thought that the bonding in MoS_2 is partially ionic. Nevertheless, considering that the A exciton for the other group VI transition metal dichalcogenides (see next section) shows the same type of response under an applied electric field irrespective of the structure of the lattice, one may be inclined to believe that the EM signal is truly associated with the exciton itself. The change of $\Delta R/R$ of about 10^{-3} for the applied field of 5×10^2 V cm^{-1} is not hard to understand, as the ionization field for the discrete states goes as n^{-4} and hence

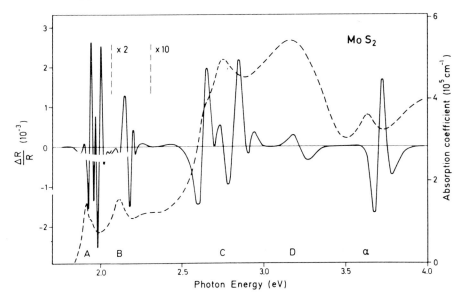

Fig. 40. Electroreflectance spectrum of 2H—MoS$_2$ at 77 K obtained with the use of the transverse method for an applied field of 500 V cm^{-1}.

higher order states should yield a large signal even for low fields, which is consistent with the fact that the ER signal in Figure 40 shows the main response in the region of higher order states and not around $n = 1$ as in the case of EA results obtained with the longitudinal method (Figure 39), (although in this case the electric field is likely to be overestimated). Besides one has to bear in mind that MoS$_2$ shows a considerable degree of anisotropy in its electrical properties. The conductivities perpendicular and parallel to the layers ($\sigma\perp$ and $\alpha\parallel$) are very anisotropic ($\sigma\parallel/\sigma\perp \sim 10^3$) which even in the case of thin crystals, would make the electric field applied in the transverse configuration non-uniform at the surface, having a component normal to it. Such an effect considered with a space charge barrier could lead to a distortion in the acting field and thus a signal at the fundamental frequency of the applied field. When the crystal is cooled down, a more symmetric type of perturbation may arise thus reducing the size of the EM signal at the main frequency. Obviously, the use of square wave modulation and biasing studies along the lines described in Section 5, would throw some light on the validity of this explanation.

10. Electroabsorption on the A Series of 3R—MoS$_2$, 3R—WS$_2$, 2H—WSe$_2$, 2H—MoS$_2$

The EA spectra shown in Figure 41 were obtained from a 1500 Å thick crystal of 3R—MoS$_2$ [30]. From biasing studies the existence of a large internal field (F_i) that could never be completely compensated was inferred. One could never

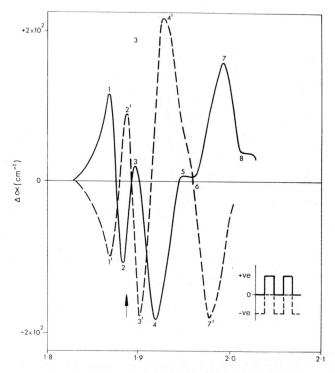

Fig. 41. Electroabsorption spectra of the A exciton of 3R—MoS$_2$ at 80 K. $F_b = 0$ (full line) and $F_b = -F_A$ (dashed line).

achieve modulation from the flat band position in any of the crystals of 3R—MoS$_2$ that were tried, but nevertheless the penetration depth of the applied field (F_A) was large enough to give all the oscillations running into the region of the continuum. The energy position of the exciton ground state ($n = 1$) is indicated by an arrow and was obtained by simultaneous measurements of transmission; that of $n = 2$ by reflectivity measurements, as it could be observed in transmission.

The full line of Figure 41 corresponds to:

$$\Delta \alpha = \alpha(F_i + 2.1 \times 10^5 \text{ V cm}^{-1}) - \alpha(F_i).$$

The peaks in this EA spectrum are numbered in order of increasing energy, in an attempt to correlate them with those found for 2H—MoS$_2$ (see Figure 39 and ref. 56). The oscillation marked 1, 2, 3 corresponds to the effect of the electric field on $n = 1$ and corresponds to its quenching plus shifting to lower energies. The oscillation 3, 4, 5 corresponds to the quenching of $n = 2$, while 6, 7 and 8 are associated with the effect of the electric field on higher order states and the continuum. At higher energies the spectrum then slightly overlaps with the EA from the B exciton which will be discussed in Section 4.

When a negative d.c. bias was applied to the a.c. electric field so as to make the latter describe negative excursions from zero (as shown in the insert), in other words, when modulation was achieved against the built-in field, reducing it, the EA spectrum shown by the broken line of Figure 41 was obtained. The effect of modulating under this condition is, as expected, to increase the absorption at the top of the exciton peaks and sharpen up the exciton states. As the broken line spectrum corresponds to:

$$\Delta\alpha = \alpha(F_i) - \alpha(F_i - 2.1 \times 10^5 \text{ V cm}^{-1})$$

the change of sign can be easily understood. The sharpening of the structure causes peak 3' to grow bigger relative to 3 and the whole EA spectrum is less broad, extending over a slightly smaller energy range. Maybe it should be added that when sine wave modulation was tried the signal at the fundamental frequency S_f looked very much like the one obtained for square wave modulation, while S_{2f} was practically undetectable, again indicating the presence of a very large internal field. (The reader is referred back to Section 5 for a detailed discussion of these effects.)

Although modulation from the flat band position could not be obtained for 3R—MoS_2 this was not the case for 3R—WS_2. Crystals of WS_2 with a small enough internal field to enable modulation from the flat band position could be grown. But, there, one has to bear in mind that differences in the penetration depth of the applied electric fields can arise by modulating either from flat band to an accumulation layer or to a depletion layer. The EA line shapes obtained for WS_2 are chosen to illustrate that only when the ideal requirement of modulation with large penetration depth is achieved, can fully meaningful line shapes be obtained.

In Figure 42 we show several EA spectra obtained for WS_2 from two crystals. Figure 42(a) and (b) correspond to the signals obtained from a 1500 Å thick crystal of WS_2 for modulating fields as indicated in the insert. They represent:

$$\Delta\alpha = \alpha(1.2 \times 10^5 \text{ V cm}^{-1}) - \alpha(0).$$

In Figure 42(a) the spectrum is reduced to one large negative peak 2 with two positive satellites 1 and 3. The signal corresponds to a quenching and shifting of the $n = 1$ state to lower energies, as one can infer from its shape in the region of the exciton ground state, which was located, as usual, from simultaneous measurements of transmission.

In Figure 42(b), besides the oscillation 1, 2, 3 associated with $n = 1$, which again, corresponds to a shift towards lower energies, the signal obtained shows oscillations 3, 4, 5, 6 in the region of higher order states and the continuum. The appearance of the oscillations 3, 4, 5, 6 in Figure 42(b) can be attributed to the fact that modulation was achieved from flat band to a depletion layer and therefore the electric field penetrated deeper into the sample than in Figure 42(a), for which the signal associated with higher order states is absent because of their large spatial extension relative to that of the electric field. However although the spectrum in Figure 42(b) shows structure in the region of higher order states and

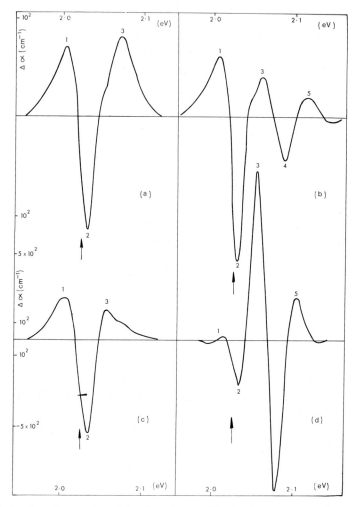

Fig. 42. Electroabsorption spectra of the A exciton of WS_2 for the several experimental situations described in the text.

continuum, it is not as prominent as that found, for instance, for 2H—MoS_2 or for 3R—MoS_2. One concludes that this was due to the electric field having what one could call an intermediate penetration depth, enough to give a signal in the higher order and continuum region but not enough to make it as prominent as in MoS_2. Nevertheless, it proved possible to achieve the same type of response. In Figure 42(c) and (d) there is shown another example of EA spectra obtained from a 500 Å thick crystal of WS_2. Figure 42(c) shows:

$$\Delta\alpha = \alpha(1.2 \times 10^5 \text{ V cm}^{-1}) - \alpha(0)$$

with an applied field oscillating describing positive excursions. The signal is again one large negative well 2 with positive satellites 1 and 3 at each side and, although a minute oscillation shows at the high energy side of peak 3, the signal is not anywhere near as prominent as in Figure 42(d) that was obtained with the same absolute value of the applied field but a different polarity. The appearance of the large signal on the higher energy side of $n=1$ means that in this case a much larger penetration depth was obtained, approximating the ideal case.

The same type of effect was found in the case of WSe$_2$ and MoSe$_2$. These again showed the existence of internal fields, which could be compensated by appropriate biasing in some samples. When modulation was achieved from a depletion layer the dominant signal was in the region of the continuum.

In Figure 43 the EA spectra of the exciton of 2H—MoS$_2$, 3R—MoS$_2$, 3R—WS$_2$, 2H—WSe$_2$ and 2H—MoSe$_2$ as obtained under conditions approximating the

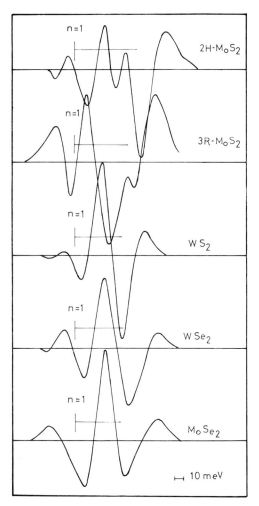

Fig. 43. Comparison of the electroabsorption spectra yielded by the *A* exciton of 2H—MoS$_2$, 3R—MoS$_2$, 3R—WS$_2$, 2H—WSe$_2$ and 2H—MoSe$_2$ at 80 k.

ideal one are shown for an applied field $F_A = 1.2 \times 10^5$ V cm^{-1}, although in the case of 3R—MoS$_2$ and 2H—WSe$_2$ no samples in which the internal field could be compensated were obtained. The photon energy scale has been shifted in order to place the position of the $n = 1$ state of all the excitons in the same vertical line. The most prominent signal for this value of the electric field appeared in the region of the continuum, which seems to be of more importance than modulation from the flat band position, although the latter requirement is necessary if a quantitative fit with the theory is attempted.

What one can learn from the spectra shown in Figure 43 is that, when dealing with excitons of large binding energies and small radii (F_I large), it is possible to separate structure arising from the ground state, higher order states and continuum, with the largest response appearing in the region of the continuum. If we take a hypothetical exciton of binding energy 50 meV in a medium with dielectric constant 7 (this exciton woud closely reflect the A exciton series for the materials that we are dealing with in this paper) it would have $F_I = 2.5 \times 10^5$ V cm^{-1}. The applied fields in Figure 43 are estimated to be 1.2×10^5 V cm^{-1} but no correction for possible screening by surface states was considered. Very crude considerations indicate that the fields can be overestimated by as much as a factor of 10, so $F_A \simeq 4 \times 10^4$ V cm^{-1}. Consequently, $F_A/F_I \simeq 0.16$. The theory [5, 6] predicts that the $n = 1$ state should shift to lower energies in agreement with the experimental findings.

For the field regime $F_A < 0.1 F_I$ it seems that the effect is to start building oscillations in the region of the continuum, giving rise to what one could call a 'band edge' signal (although, as previously stated, 'band edge' is only a meaningful term if one works on the framework of a one-electron model). In the next section it will be shown how the signal develops towards one single large oscillation when the signal from $n = 1$ takes over, masking the fine structure. There is another obvious factor that is critical for the detection of fine structure in an exciton series, and this is that the experiment has to be done under such conditions that the broadening (Γ) of the exciton line is not too large relative to the exciton Rydberg R.

As it was shown in the case of PbI$_2$, no fine structure can be detected when $\Gamma \simeq R$. Consequently, it is a combination of small enough values of F_A relative to F_I, Γ relative to R, and large penetration depths of F_I that enable the separation of fine structure in exciton series. The lack of structure around $n = 2$ for the case WS$_2$, WSe$_2$ and MoSe$_2$ is due to the fact that this line is broader for these excitons than it is for 2H—MoS$_2$ and 3R—MoS$_2$. Nevertheless, if one takes the 'band edge' signal as a marker for the onset of the continuum, then the lines drawn from the $n = 1$ energy position in Figure 43 indicate the estimated binding energies from the EA spectra line shapes, which are in very fair agreement with those from transmission spectra obtained at very low temperatures [62] after corrections for size effect [57] are taken into account.

11. Development of the EA Line Shapes from the Low Field to the High Field Regime

The development of the EA for 3R—MoS$_2$ for increasing fields is shown in Figure 44 [48]. Figure 48(a) shows the EA spectrum for $F_A = 4 \times 10^4$ V cm^{-1}. The separation of the several signals in the spectrum is very clear; peaks 1, 2, 3 arise from the effect on $n = 1$; peaks 3, 4, 5 on $n = 2$ and peaks 5, 6, 7 on higher order states and the continuum. As the field is increased (Figure 48(b), (c) and (d)) the fine structure disappears and for an applied field of 6.5×10^5 (Figure 48(d)) the structure is highly quenched, the several peaks having mixed and overlapped, leaving behind one single broad oscillation (1, 2, 7) whose shape is quite similar to the shape predicted theoretically in the high field regime [5] and found experimentally in materials with excitons of smaller Rydbergs [44, 48].

Figure 48 shows that if signals under small enough electric fields could be detected one finds that discrete states with larger binding energies will remain relatively unaffected. Increasing the magnitude of the field will lead progressively to effects on the $n = 3$, $n = 2$ and eventually $n = 1$, higher order states having broadened into the continuum. Although this prediction has not been proved

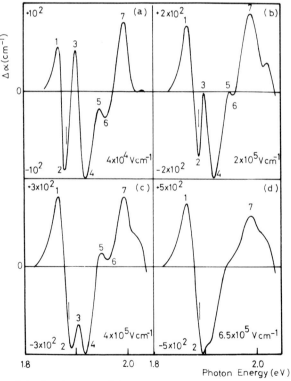

Fig. 44. Illustration of the development of the electroabsorption response from the low field regime to the high field regime as obtained at 80 K for the A exciton of 3R—MoS$_2$.

explicitly by theory, it seems almost self-evident and experimental evidence in Cu_2O seems to support this pattern of behaviour [63]. Observe that in Figure 44(d) the signal still corresponds to a shift of $n = 1$ to lower energies, which indicates that according to theory, $F_A < F_I$.

12. Summary of Electromodulation Experimental Results

The main conclusions from the EM experiments reported in this section can be summarized as follows:

(i) It is important to perform the measurements in such conditions that the applied electric field penetrates as uniformly as possible if a spectrum showing meaningful structure is to be achieved.

(ii) The existence of internal fields can make modulation from the flat band position difficult to achieve, or occasionally impossible. Although this can be a problem if a quantitative comparison with theory was intended, it seems to be a less important problem as far as the detectability of fine structure is concerned than is incomplete penetration depth of the electric field, as suggested by the cases of $3R-MoS_2$ and WSe_2, where modulation from the flat band position could not be achieved. One can always modulate against this internal field improving the changes of detecting fine structure, as in the case of the sharpening of the spectrum of the A exciton of $3R-MoS_2$ when the internal field was reduced by bias.

(iii) The use of unbiased sine wave modulation can be very misleading as far as the real effect of the electric field is concerned, as the existence of an internal field or the distortion induced by a space charge layer on one half of the cycle, even when modulation is achieved from the flat band position, can produce a very unrealistic EA spectrum, yielding signals at the fundamental frequency (f) as well as at harmonics.

(iv) Screening of the applied electric field by surface states almost certainly plays a role in the kind of experimental arrangement used in these measurements and the quoted values of the applied electric field are overestimated.

(v) The smaller Γ/R ratio one can obtain, the better chance one has to separate distinct structure associated with the exciton ground state and higher order states, in which case, good estimates of the exciton binding energy, and consequently other parameters, can be obtained from the shape of the EA signal. We attribute the detection of richer structure in the case of $2H-MoS_2$ and $3R-MoS_2$ to the fact that they had the lowest Γ/R relation of all the materials studied.

(vi) Qualitatively at least these excitons behave as predicted theoretically. In the field regime studied this is when $0.85 F_I > F_A$, the exciton ground state is predicted to undergo a shift to lower energy. The EA shapes obtained all imply a shift to lower energies in agreement with theory.

(vii) Higher order states, because of their smaller binding energies and bigger radii, are more affected by small applied fields than the exciton ground state. For increasing fields then, the effect arising from $n=1$ takes over, the fine structure is lost and the EA signal develops towards a large negative well near the energy position of $n = 1$ (Figure 44(d)) with two positive satellite peaks at each side.

(viii) Although the anisotropic nature of the excitonic transitions considered here has not been explicitly studied, it should be taken into account if a quantitative comparison with theory is to be undertaken. Modulation experiments with different geometries should certainly give some information about the anisotropy of the excitonic parameters. ($F_A \perp c$-axis as well as $F_A \| c$-axis).

13. Electromodulation Results for Photon Energies Above the Fundamental Absorption Edge

The interpretation of the EM spectra in this energy region is understood less than at the fundamental absorption edge. Some contradictory statements as to whether one-electron theories or theories including the electron-hole interaction have to be used to explain the experimental data are often encountered in the literature.

Nevertheless because of the selectivity of EM techniques in yielding structure at critical points in the optical spectra, thus eliminating the structureless background which appears in conventional absorption or reflectivity measurements, EM techniques have fruitfully been used to detect structure that would otherwise have been very difficult to resolve.

It is, perhaps, for the layer material GeSe that the most determined effort has been made to reproduce theoretically the experimental results, but in most other cases only a qualitative interpretation has been attempted.

13.1 ELECTROMODULATION EXPERIMENTS ON GaSe ABOVE THE FUNDAMENTAL ABSORPTION EDGE

Figure 45 shows the ER spectrum at room temperature in the photon energy range from 1.9 to 5.3 eV and for $F \| c$-axis [64]. In the energy region around 2 eV the EA spectrum is shown as well. The structure in this region is associated with the exciton at the fundamental absorption edge and the authors are in close agreement with the estimates of low binding energy obtained in Section 10, where this exciton was discussed. The full line spectrum above 3 eV corresponds to the experimental results and the broken line corresponds to a theoretical fit attempted by the authors, who neglected the electron-hole interaction and used broadened electro-optical functions for their theoretical fitting. The observed structure was interpreted as due to M_1 and M_2 three dimensional singularities. Attributing E_1 and $E_1 + \Delta E_1$ to transitions from a spin-orbit split valence band formed by the p states of Se at an M_1 critical point, while E_2 was explained as originating at a saddle point in the (0 0 1) direction from an M_1 edge. These assignments were made by considering the signs of the reduced masses found along the k_z direction and the value of the spin-orbit splitting obtained from the theoretical fit, while for the same reasons E_3 and E_4 were attributed to M_2 type of critical points. As one

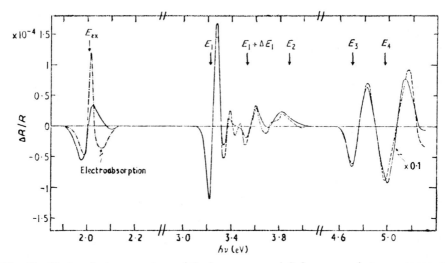

Fig. 45. Electroreflectance spectrum of the layer compound GaSe, measured at room temperature and near normal incidence. Around the edge the electroabsorption ($\Delta I_r I_r$) spectrum is shown as well. A square wave electric field is applied parallel to the c axis; the reflecting surfaces are the cleavage planes of the crystal. Broken curve, theory; full curve and chain curve, experiment.

can see from Figure 45 the theoretical fit looks rather satisfactory. Nevertheless, later ER measurements at 77 K [65] were interpreted differently again with the aid of the one-electron theory. In neither case a theoretical fit including exciton effects was attempted.

As it will be discussed in Section 21, excitonic effects have had to be considered in order to explain the temperature modulation results for the similar compound GaS at photon energies above the fundamental absorption edge. It is quite likely that the inclusion of exciton effects in the interpretation of the EM spectra in GaSe might give as good or even better fit than the one-electron theories with a more meaningful approach to the problem.

13.2. Electromodulation experiments on some layer materials for photon energies above the fundamental absorption edge

In the case of electromodulation results for some of the transitions above the fundamental absorption edge of the group VI transition metal dichalcogenides, there can be no doubt about the importance of exciton effects.

The EA spectrum ($\Delta \alpha = \alpha(F_A) - \alpha(0)$) of $2H$—MoS_2 in the energy region between 1.8 and 2.8 eV with F_A estimated to be 1.6×10^5 V cm^{-1}, under the most favourable conditions, is shown in Figure 46 [30].

The positions of the A_1, B_1, A'_{12} and C transitions are indicated by arrows. The selectivity by electromodulation spectroscopy to critical points in the electronic band structure is clearly revealed by the spectrum of Figure 46 when compared to the straight absorption spectrum of $2H$—MoS_2. In the latter B_1, A'_{12}, and C appear as features superimposed on a large structureless background, in the EA

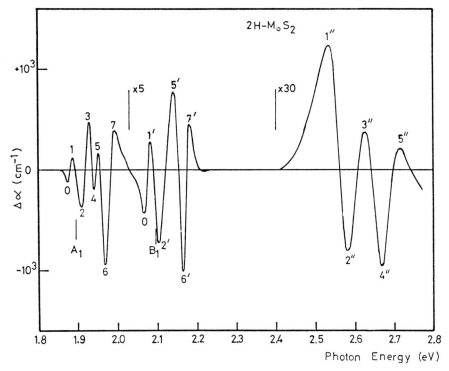

Fig. 46. Electroabsorption spectrum of 2H—MoS$_2$ as obtained with the longitudinal method of 80 K ($F_A = 1.6 \times 10^5$ V cm^{-1}).

spectrum no signal is detected in the energy region between critical points. It should be emphasized that the size of the EA signal gets smaller for higher energies (×5 for the B region, ×30 in the A'_{12} and C regions).

The EA spectrum in the B exciton region looks very much like that obtained for the A exciton (which was discussed in detail in Section 12), but with two differences:

(a) The signals are much smaller in size (the scale is ×5 for the B exciton region).

(b) The oscillation corresponding to $n = 2$ for the A exciton (peaks 3, 4, 5 in Figure 39) is not present for the B exciton.

The energy separation between B_1 and 5′, 6′, 7′ are very close to those occurring between A_1 and the corresponding peaks 5, 6, 7. The values are:

53 meV between A_1 and 5,   52 meV between B_1 and 5′,

68 meV between A_1 and 6,   72 meV between B_1 and 6′,

103 meV between A_1 and 7,   85 meV between B_1 and 7′.

The similarity of the B exciton EA spectrum with the one for the A exciton suggests that both excitons have similar binding energies. Although this might be expected if A and B are associated with transitions from a spin-orbit split valence band, it has been deduced [58, 59, 66], from careful measurements of transmission and reflectivity in which higher order states associated with the B exciton were observed, that the binding energy of the B exciton is about 130 meV. If we are to attribute similar binding energies, then we have to account for the lack of an EA signal associated with $n = 2$ of the B series.

It should be emphasized that the B exciton is much broader than the A exciton in the field free case, probably due to auto-ionization of the discrete states into the continuum associated with the A exciton. As a general rule, the broader the structure in straight absorption, the broader the EA spectrum. We therefore conclude that for the B series, the EA signal associated with $n = 1$ mixes with that from higher order states forming the single large peak 5'. Observe that the EA spectrum for the B exciton is very similar to that obtained for the A exciton of WS_2, WSe_2 and $MoSe_2$ (see Section 13). In these cases the lack of clear contributions from $n = 2$ was attributed to smaller binding energies as can be seen from an overall comparison in energy of the total signal relative to that in MoS_2.

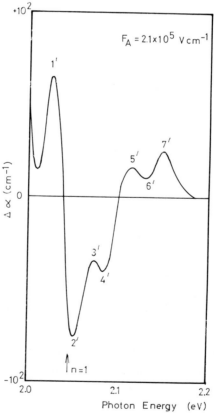

Fig. 47. Electroabsorption spectrum yielded by the B exciton of 3R—MoS_2 at 80 K ($F_A = 2.1 \times 10^5$ V cm^{-1}).

This is because the broadening in units of exciton Rydberg for these materials is bigger at the same temperature.

In the region of A'_{12} and C, the EA signal is even smaller in size than for B. The signals arising from the electric field effects on A'_{12} and C are very broad and overlap quite strongly. However it is clear from the two negative peaks occurring near the field free positions of A'_{12} and C, that both features are quenched by the field. The asymmetry of the positive peaks on either side of A'_{12} imply in addition a shift of A'_{12} to lower energies, but because of overlap with the signal from C this cannot be considered definite. Although excitons at energies well above the edge are known to occur the excitonic nature of A'_{12} and C cannot be confirmed from this study alone, nevertheless one is perhaps inclined to think of these transitions as having excitonic character because of the shape of the EM signal.

Figure 47 shows the EA spectrum, $\Delta\alpha = \alpha(F_i) - \alpha(F_i - F_A)$ for a crystal of 3R—MoS_2 [30] in the region of the B exciton. The actual value of F_i could not be obtained by any of the means used in other cases and it was certainly bigger than 3×10^5 V cm^{-1}. The spectrum shown in Figure 47 exhibits a series of overlapping features giving several peaks (1', 2', 3', 4', 5', 6', 7') which have been numbered, in an attempt to correlate them with the EA of the A exciton. Observe that the EA spectrum in Figure 47 looks very similar to the one shown in Figure 44(c) which is obtained for the A exciton for a higher F_A. This similarity is explained by the fact that the B exciton is broader than the A in the field free case, and a smaller electric field is therefore required to produce a similar broadening.

The energy separations between the exciton ground states and the several peaks in the two spectra compare well:

100 meV from 7' to B_1: 103 meV from 7 to A_1
66 meV from 5' to B_1: 60 meV from 5 to A_1
38 meV from 4' to B_1: 33 meV from 4 to A_1
24 meV from 3' to B_1: 19 meV from 3 to A_1
23 meV from 1' to B_1: 23 meV from 1 to A_1

One would conclude, as in 2H—MoS_2, that the B exciton has a binding energy very close to that of the A exciton, although in the field free case it is broader, probably due to auto-ionization into the continuum of the A exciton. This explains the smaller and broader EA signal obtained for the same F_A.

The EA spectrum $\Delta\alpha = \alpha(F_A) - \alpha(0)$ of a crystal of WS_2 in the energy region between 1.9 eV and 2.9 eV is shown in Figure 48. The spectrum associated with the A and B excitons were obtained for $F_A = 1.2 \times 10^5$ V cm^{-1}, but in the region of A' and C, the field used was $F_A = 6 \times 10^5$ V cm^{-1} as no clear signal could be detected for smaller fields. Observe that the scales are $\times 10$ in the B exciton region and $\times 5$ in the A' and C regions.

The EA spectrum of the B exciton presents a shape similar to the one found for the A exciton, but much broader. The energy position of the exciton ground state is very close to the negative peak 2', and the largest response appears at higher

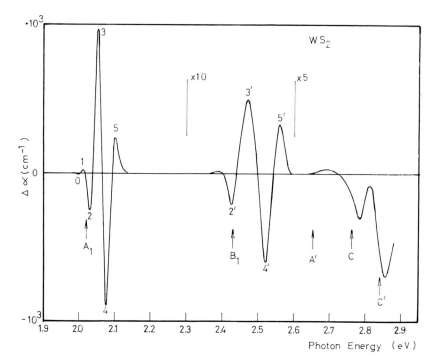

Fig. 48. Electroabsorption spectrum of WS_2 at 80 K as obtained with the longitudinal method.

energies (3′, 4′, 5′). This by itself suggests that the B exciton has a relatively large binding energy (>50 meV), but the exact value is difficult to ascertain. The fact that the EA spectrum of the B exciton is extended over a larger energy region than that for the A exciton can be taken as an indication of a larger binding energy than A (~50 meV), or perhaps simply as a consequence of the fact that B is broader in the field free case. In the case of the A exciton it was seen (Section 13) that the continuum for the A exciton falls at about the energy position of peak 4. If 4′ is taken as having a similar origin as 4, then the binding energy for B would be about 90 meV.

No EA structure related to the A′ transition, observed in absorption measurements, could be detected. In the C region the EA response obtained corresponds to a shifting to lower energies of the C transition accompanied by quenching. A larger negative peak (labelled C′) appears at about 2.855 eV with unknown origin.

The EA spectrum $\Delta\alpha = \alpha(F_i) - \alpha(F_i - F_A)$ of WSe_2 in the energy region between 1.6 eV and 2.8 eV and for $F_A = 1.2 \times 10^5$ V cm^{-1} is shown in Figure 49. The EA spectrum associated with the B exciton overlaps with that associated with the A′ exciton. The slight increase in absorption at the energy position of the B exciton ground state is considered to arise from an overlap with the EA response produced by the A′ exciton (satellite structure due to its broadening by the field).

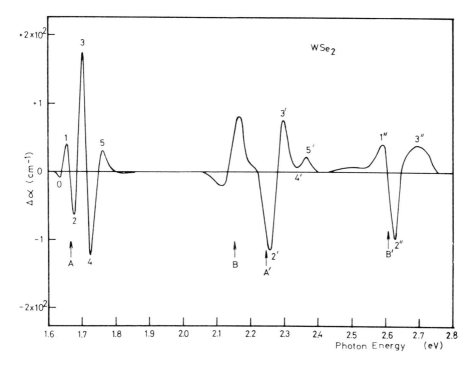

Fig. 49. Electroabsorption spectrum of WSe$_2$ at 80 K as obtained with the longitudinal method.

The EA spectrum in the region of A'_1 corresponds to a quenching of the exciton ground state. Peak 2' giving rise to the satellite peak 3'. More structure appears at slightly higher energies, namely the oscillation 3', 4', 5'. The EA spectrum for the A' exciton is broader than for the A exciton. The EA spectrum associated with the B' exciton corresponds to a quenching by the field and broadening. The satellite 3" is very broad and could easily hide finer structure.

Figure 50 shows the field induced charges in the real and imaginary parts of the dielectric constant of PbI$_2$ at 77 K, and for a field of 3.5×10^4 V cm^{-1} applied perpendicular to the c-axis [39]. These spectra were obtained from a Kramers-Kronig analysis of the ER results on PbI$_2$. All the detected structure arises from the effect of the electric field on metastable excitons and so far no theoretical attempts to interpret the line shapes obtained have been made. However, the positions of the negative minima in $\Delta \varepsilon_2$ fall at about the photon energy positon of the peak of the exciton in ε_2, in a similar manner to the behaviour shown by the exciton at the fundamental absorption edge (see Section 8). In the case of PbI$_2$ the EM results certainly should be interpreted by theories including excitonic effects.

It is obvious that a good deal of theoretical work is needed if the EM structure above the fundamental edge found on these layer materials is to be understood. It certainly seems that the theory has to include exciton effects, especially in the more ionic type of layer material like PbI$_2$.

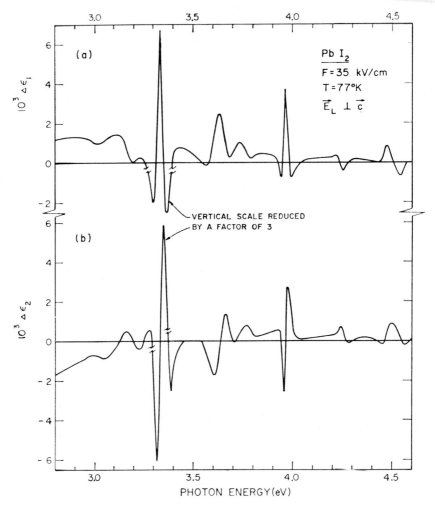

Fig. 50. Field induced charge of (a) real and (b) imaginary parts of the dielectric constants of PbI_2 at 77 K as obtained from a Kramers-Kronig analysis of the electroreflectance spectrum obtained with the transverse method.

14. EM Experiments on Very Anisotropic Layer Materials – As_2Se_3

In the experimental results described so far the materials studied were uniaxial. In some layer materials (As_2Se_3, As_2S_3, Bi_2Se_3, Bi_2Te_3 among others) the crystals are biaxial. EA experiments on As_2Se_3 [30] have revealed the anisotropy of the exciton parameters and given an indication of the power of EM techniques to pick up effects associated with the anisotropy as well as detecting fine structure.

As_2Se_3 is a low symmetry layer material (monoclinic). Each As atom is surrounded by three nearest-neighbour Se atoms and each Se atom is surrounded by two nearest-neighbour As atoms. The layers are loosely bonded by weak Van der Waal's forces and within the layers the As_2Se_3 molecules form spiral chains parallel to the c-axis in a manner similar to Se.

The symmetry within a layer is essentially orthorhombic as the c-axis is at almost 90° to the c-axis, while the b-axis is perpendicular to the layers. Consequently there are two principal directions of polarization in the plane of the layers which yield quite different optical spectra.

As As_2Se_3 has four molecules per unit cell, 24 atoms, the Brillouin zone is small and the electronic energy bands are probably fairly flat and with many transitions allowed over a short photon energy interval. This explains the very rich structure near the fundamental absorption edge which is found experimentally [67]. The complexity of the crystal symmetry makes a band structure calculation a very difficult task and as yet there is not a satisfactory one available.

Far infra-red and Raman studies of As_2Se_3 crystals [68] have shown that it is the diperiodic symmetry of the individual layer and not the usual triperiodic crystal symmetry which dominates the lattice spectra. This leads to the idea of considering As_2Se_3 as a molecular crystal in which the molecular unit is infinitely extended in two dimensions in the plane of the layer. This kind of asymmetry should show itself in the electron excitations as well as in the lattice vibrations. To investigate this point absorption and electroabsorption measurement were performed on As_2Se_3 single crystals [30]. The light polarization vector was oriented parallel to the c-axis and parallel to the a-axis as was the applied electric field, allowing a detailed study of the anisotropy of the optical spectra.

The EA spectra were obtained using a gap like configuration (transverse method) and no sign of the presence of internal fields were detected, consequently the field induced charge in the absorption coefficient $\Delta\alpha$ truly corresponds to the difference in absorption when the electric field is applied ($\alpha(F)$) and when it is not ($\alpha(0)$).

In Figure 51 the A spectra of single As_2Se_3 crystals at 80 K and 10 K for $E \perp c$-axis and $E \| c$-axis are shown. The spectrum for $E \perp c$ is very rich in structure especially at 10 K, where a series of transitions marked a, b, c, d, e and f were detected, while for $E \| c$ no structure can be clearly seen. These transitions are attributed to exciton formation and accordingly they sharpen up considerably at low temperature.

Because of the extreme anisotropy of the As_2Se_2 crystal structure the effective masses of the exciton states and the dielectric constant of the medium must have a tensorial nature. Thus the exciton states will not be isotropic and the Wannier-Mott description of excitons is not valid. Neither also is the Frenkel descriptions correct; although localized in the layer plane, the excitons will be extended in the layers. In fact they have to be considered highly anisotropic excitons with a nature somewhere between the Wannier and the Frenkel type.

Figures 52 and 53 show the EA spectra for $E \perp c, F \| c$; $E \perp c, F \perp c$; $E \| c, F \| c$ and $E \perp c, F \perp c$ and $E \perp b$ in all cases at 80 K, along with the field dependence of the heights of the respective EA peaks. The EA spectrum in Figure 52(a) ($E \perp c$,

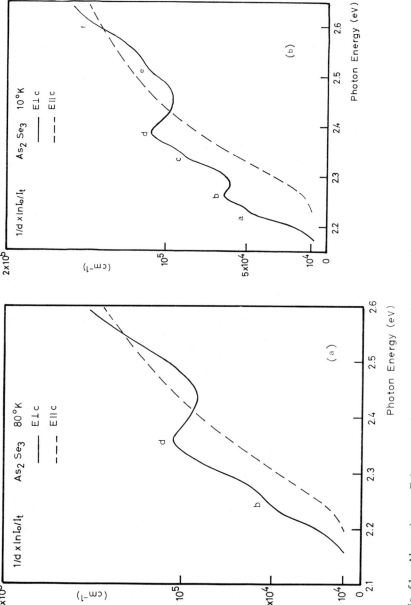

Fig. 51. Absorption coefficient as As_2Se_3 at (a) 80 K and (b) 10 K for $E \perp c$ (full line) and $E \parallel c$ (dashed line).

$F \| c$) is very rich in structure. The oscillation 1, 2, 3 arises from the a exciton; 3, 4, 5 from the b; 5, 6, 7 from c; 7, 8, 9 from d and 9, 10 from e. Peaks 2, 6, 8, 10 and 12 are due to the field quenching of the excitonic transitions and fall at the energy position of the exciton peak respectively, and peaks 1, 3, 5, 7, 9 and 11 are due to the satellites created by the field broadening of the several transitions. Due to the proximity of the several transition, there is some overlap of the EA signals leading to common satellite peaks.

Figure 52(b) shows the EA spectra for $E \perp c$, $F \perp c$. The spectra are very similar to the ones of Figure 52(a), and again peaks 2, 4, 6, 10 and 12 correspond to the electric field quenching of the excitonic transitions a, b, c, d and e respectively; the EA signal associated with a and b this time shows a common satellite peak 3. Nevertheless the size of the EA signals is smaller for the same applied field for $F \perp c$ than for $F \| c$, as can be seen from Figure 53(a) and Figure 53(b). The reduction in size is about a factor of 3.4 for 4; 4.5 for 8; 4.88 for 10 and 3.14 for 12. The ratio for 2 cannot be estimated easily because of heavy overlap between the signals for $F \| c$.

Figure 53(a) and 53(b) show the EA spectra for $E \| c$ with $F \| c$ and $F \perp c$ respectively. Only structure associated with three transitions c (oscillation 1, 2, 3); d (oscillation 3, 4, 5) and e (peak 6) are detected for $F \| c$. For $F \perp c$ the signal is very much smaller and the field dependence could not be followed. Besides no signal associated with e could be detected, presumably because it was buried in noise.

The ratio of the sizes for the same applied electric fields are 3.3 for peak 2 and 2 for peak 4, although in the case of peak 4 overlapping with neighbouring structures is important and the size ratio might be wrong.

From the measured ratios of the size of the signals an estimate of the anisotropy of the exciton may be obtained. For excitonic transitions the size of the signal h follows a power law where: $h = (F/F_I)^n$. n is an exponent and F_I is the ionization field of the exciton. For this case of As_2Se_3 $n = 1.4$.

For $F \| c$ we then have, for the case of the c transition, for instance:

$$\frac{h^2}{h'^2} = 3.4 = \left(\frac{F/F_I}{F/F_I'}\right)^{1\cdot 4} \tag{36}$$

then:

$$F_I' = 2.39 F_I$$

consequently:

$$\frac{R}{a_\perp} = 2.39 \frac{R}{a_\|} \quad \text{and} \quad a_\| = 2.39 a_\perp$$

leading to the result that the exciton is more extended along the c-axis, this is to say along the molecular chains parallel to the c-axis.

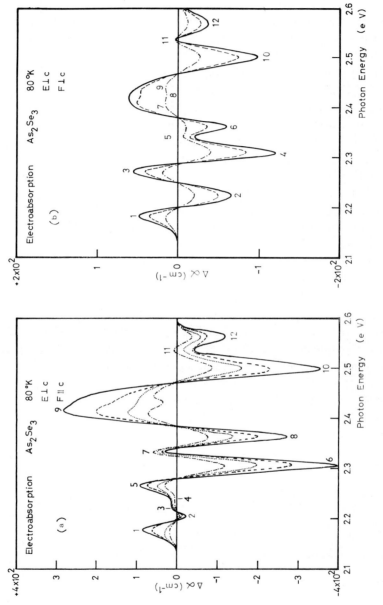

Fig. 52. Electroabsorption of As_2Se_3 at 80 K (a) $E \perp c$ and $F \perp c$ for increasing electric fields ($2.6 \times 10^4 \, V \, cm^{-1}$, $5.3 \times 10^4 \, V \, cm^{-1}$, $6.6 \times 10^4 \, V \, cm^{-1}$).

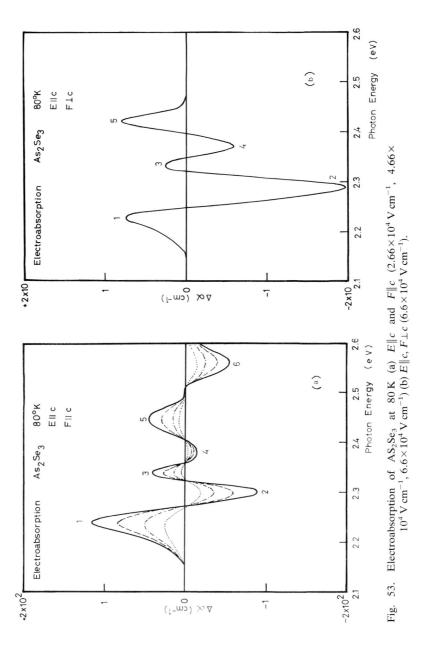

Fig. 53. Electroabsorption of As_2Se_3 at 80 K (a) $E\|c$ and $F\|c$ (2.66×10^4 V cm^{-1}, 4.66×10^4 V cm^{-1}, 6.6×10^4 V cm^{-1}) (b) $E\|c$, $F\perp c$ (6.6×10^4 V cm^{-1}).

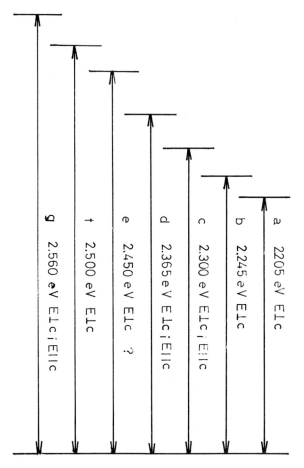

Fig. 54. En energy levels scheme for AS_2Se_3 as obtained from electro-absorption measurements.

Figure 54 gives the energy and polarization dependence of the detected transitions. The final interpretation of the polarization behaviour of the observed peaks and the energy of the excitations must await a band structure calculation for As_2Se_3.

In order to get an accurate idea of how localized within the layer these excitonic transitions are, one should perform electromodulation experiments with the applied field along the b-axis and compare the size of the response with the other possible orientations of the field. These experiments are currently under way [44].

15. Thermomodulation Techniques

A usual method for thermomodulation is to attach the sample to a heat sink and pass an alternating current through it. Where the resistivity of the sample is too high to allow a sizeable enough current to be drawn, it is convenient to put the sample in intimate contact with an electrically conducting substrate and modulate its temperature by passing an alternating current through the subtrate. In this case both sample and subtrate have to be attached to a heat sink. A method that works quite well and allows one to perform thermotransmission experiments [56] is to use a sapphire subtrate covered with a semitransparent or transparent metal film (SnO_2, nickel or some other suitable metal), on top of which the sample is attached. The alternating current is passed through the metallic film indirectly heating the sample.

If one assumes that the sample temperature is uniform and that the increase in the average temperature of the sample $\langle \mathfrak{T} \rangle$ above that of the heat sink, $\mathfrak{T}_0$ is much larger than the amplitude $\Delta \mathfrak{T}$ of the temperature modulation. Then the modulation efficiency γ, defined as the ratio $\Delta \mathfrak{T}/\langle \mathfrak{T} \rangle - \langle \mathfrak{T}_0 \rangle$, for square pulses, is given by [1]:

$$\gamma = \frac{\Delta \mathfrak{T}}{\langle \mathfrak{T} \rangle - \langle \mathfrak{T}_0 \rangle} = 0.405 \frac{Q\tau}{c}, \tag{37}$$

where c is the heat capacitance of the sample, τ the period of the modulation and Q the heat leak per unit time and unit temperature difference between sample and sink.

For sine wave modulations with a d.c. bias equal to the root-mean-square of the modulating current:

$$\gamma = 0.22 \frac{Q\tau}{c}. \tag{38}$$

From (37) and (38) one can see that the efficiency is almost double for square wave pulses and that the efficiency decreases for increasing frequencies. In addition one must keep the heat capacity c small and the heat leak Q large. By using a sample as thin as possible one gets a better heat sink Q. However the thickness of the sample is determined by what kind of optical measurements one wants to take; for transmission measurements the thicknesses required are less than for reflection measurements. As most layer materials can be easily thinned down, one can usually perform thermotransmission measurements attaining a good modulating efficiency.

One of the problems associated with thermomodulation is that it is usually very difficult to measure $\Delta \mathfrak{T}$, making absolute measurements of $\Delta T/T$ or $\Delta R/R$ difficult.

An increase in temperature has two effects in an optical transition in an insulator or semiconductor, one is to shift the energy of the critical point and the other is associated with a temperature induced broadening. In thermomodulation experiments the detected change in transmission (ΔT) or reflectivity (ΔR) is:

$$\Delta T = T(\mathfrak{T} + \Delta \mathfrak{T}) - T(\mathfrak{T})$$

or

$$\Delta R = R(\mathfrak{T} + \Delta\mathfrak{T}) - R(\mathfrak{T})$$

For a small temperature modulation $\Delta\mathfrak{T}$, one essentially measures the derivative of T or R with respect to temperature.

Assuming that T can be approximated by the expression $T = e^{-\alpha d}$ where α is the absorption coefficient and d the thickness of the sample, one can write:

$$\frac{\Delta T}{T} = -d\left[\frac{\partial\alpha}{\partial\omega_g}\frac{\partial\omega_g}{\partial\mathfrak{T}} + \frac{\partial\alpha}{\partial\Gamma}\frac{\partial\Gamma}{\partial\mathfrak{T}}\right]\Delta\mathfrak{T}, \tag{39}$$

where Γ is the broadening parameter and the differentials have been replaced by incremental changes. A similar expression for reflectivity is:

$$\Delta R = \left[\frac{\partial R}{\partial\omega_g}\frac{\partial\omega_g}{\partial\mathfrak{T}} + \frac{\partial R}{\partial\Gamma}\frac{\partial\Gamma}{\partial\mathfrak{T}}\right]\Delta\mathfrak{T}. \tag{40}$$

In expressions (39) and (40), the quantity $\partial\omega_g/\partial\mathfrak{T}$ is usually negative with typical values of about -5×10^{-4} eV K^{-1}, although it can be positive in some cases, for instance in the fundamental edge of the lead dichalcogenides. The term $\partial\Gamma/\partial\mathfrak{T}$ is usually much smaller (typically 10^{-4} eV K^{-1}) and, for this reason, the thermomodulated signals usually look very much like a pure 1st derivative of the optical spectrum with respect to photon energies [24]; in fact, the signals yielded, except for scaling factors related to the temperature coefficients of the gap, are very much like those obtained from wavelength modulation, although sometimes with opposite sign. In principle therefore, thermomodulation does not offer much more than the possibility of obtaining first derivative like signals, which, of course, improves the possibility of detecting faint structure. Nevertheless in the case of layer materials, which very often can be considered as two dimensional lattices, it has been suggested [69] that thermomodulation can be a very useful technique to determine the type of critical point involved in a transition. As it is well known by neglecting the layer–layer interaction and consequently assuming a two dimensional lattice, the only available critical points are those of the M_0 and M_1 type, which show as a step and a logarithmic singularity respectively.

In this case there is experimental evidence that the main effect of temperature is to displace the energy of the M_0 singularity and broaden the M_1 singularity [69] in agreement with theoretical calculations [70]. As it will be shown in the next section on experimental results, quite good fits are obtained with the theory in the case of graphite. Nevertheless the necessity of incorporating exciton effects in the theory was necessary in the case of GaS and GaSe [71, 72, 73].

15.1. THERMOMODULATION EXPERIMENTS ON SOME LAYER MATERIALS
(GaSe, GaTe, Graphite, GaS)

Thermomodulation experiments on the exciton at the fundamental absorption edge of GaSe have been carried out [71] and the changes of the real and

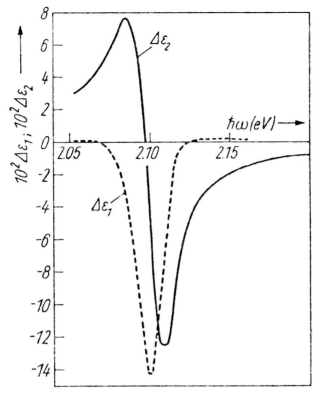

Fig. 55. Relative changes of the real and imaginary parts of the dielectric constant of GaSe as derived from the ER data by application of the Kramers-Kronig dispersion relations.

imaginary parts of the dielectric constants as derived from a Kramers-Kronig analysis of the thermoreflectance data are shown in Figure 55. A phenomenological interpretation of the data in Figure 55 was attempted by fitting a Lorentzian line shape to the exciton line and calculating the derivative like signal produced by a shift in the energy position of the exciton. The authors estimated accurately the energy position of the exciton ground state at liquid nitrogen temperature to be at 2.093 eV with a halfwidth of 41 meV.

The purely derivative nature of the signals was verified and its lower capability to resolve neighbouring structures compared with electromodulation was acknowledged by the authors. Nevertheless, thermomodulation has been put to very good use to study the anisotropy of the optical absorption of GaSe of the exciton at the fundamental absorption edge [72] by measuring thermoreflectance with polarized light from a face of a GaSe crystal normal to the plane of the layer. In this way it was possible to show that the direct exciton transition is very strong for $\bar{E} \| \bar{c}$-axis while it is twenty times weaker for $\bar{E} \perp \bar{c}$-axis. These results showed that the available band structure [74, 75] calculations failed to give a correct description of the exciton transitions. A recent analysis of selection rules of this transition [50] provides a description for these thermomodulation results.

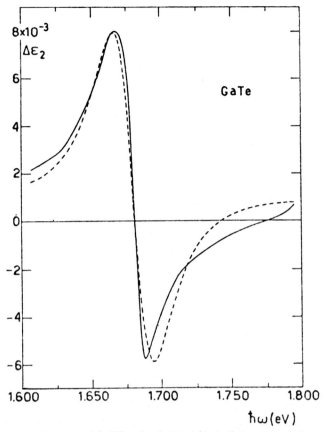

Fig. 56. Comparison of measured (solid) and calculated (dashed) line shapes of $\Delta\varepsilon_2$ in the exciton region of GaTe. The best fit has been obtained at 284 K taking $E_0 = 1.676$ eV and $\Gamma = 23.5$ meV.

Thermoreflectance experiments on the absorption at the fundamental absorption edge of GaTe [76] have been used as well in order to determine accurately the energy position and halfwidth of the exciton. Figure 56 gives a comparison of the measured (solid) and calculated (dashed) line shapes of ΔE_2. The measured ΔE_2 was obtained from a Kramers-Kronig analysis of the thermoreflectance data. As in the case of the exciton in GaSe [71] the best fit was obtained by assuming a rigid shift of the whole structure to lower energies, reiterating the purely derivative nature of the signals yielded by thermomodulation techniques in the region of the fundamental absorption edge. From the analysis of the thermoreflectance data at 284 K, the authors assigned the position of the exciton at 1.676 eV with a halfwidth of 23.5 meV.

With this type of thermomodulation experiment, in principle one should be able to determine the dependence of the exciton halfwidth as a function of temperature with great accuracy by performing the experiments at different temperatures and obtain estimates of the binding energy of the exciton involved by comparing the results with the theoretically broadened exciton lines predicted by Elliot's theory (see Figure 2).

Strong polarization dependence of the thermoreflectance spectrum of GaTe between 1.6 and 4.75 eV have been measured [77] which should be taken as a result of the low symmetry of the GaTe lattice. Because of the complexity of the band structure of GaTe it is difficult to analyse the spectra obtained.

However, thermomodulation seems to provide the possibility of obtaining extra information about the type of critical point involved in an optical transition, when one is dealing with essentially two-dimensional materials.

In a two-dimensional model one can only have M_0 and M_1 type of singularities, which give rise to step and logarithmic singularities, respectively. In two dimensions the effect of temperature on an M_0 singularity is to shift the structure, while for an M_1 it is to broaden it, consequently one would expect a different type of shape for their thermomodulation spectra and an unambiguous assignment as a result.

This approach has been used to interpret the thermoreflectance spectrum of graphite [69] and the authors were able to verify the M_0 and M_1 nature of the transition occurring in graphite at around 5.96 eV and 5.1 eV respectively. The experimental data and theoretical fit are shown in Figure 57. No inclusion of

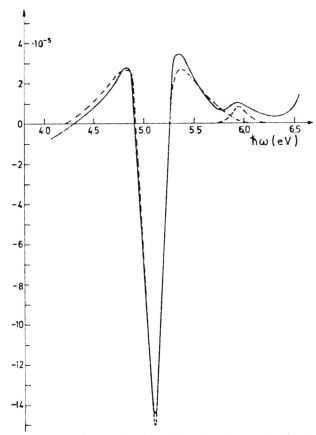

Fig. 57. Comparison between the experimental (solid lines) and theoretical (dashed lines) change of ε_2 due to an M_1 saddle point located at 5.11 eV and an M_0 threshold at 5.96 eV.

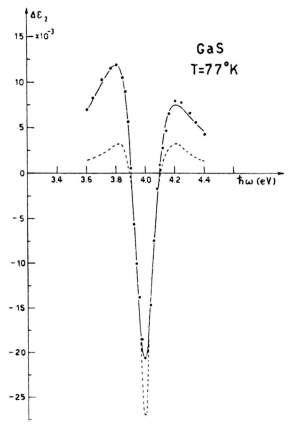

Fig. 58. A theoretical fit to the thermomodulation spectrum of GaS at 77 K, with exciton effects excluded (dashed line) and with exciton effects included (line of points).

exciton effects was attempted in this work. Nevertheless, the incorporation of exciton effects was necessary to explain the thermoreflectance of GaSe and GaS [71, 72, 73].

When an attempt was made to interpret the thermoreflectance spectrum with the aid of one electron models the agreement obtained was poor, but very good agreement could be obtained when exciton effects were included. An illustration is given in Figure 58 for the case of GaS [71].

As in the case of electromodulation one can conclude that in thermomodulation exciton effects are very important even in the energy regions well above the fundamental absorption edge.

16. Final Remarks

Among the recent modulation work that has not been discussed one should mention a recent attempt to interpret the effect of an electric field on the bound states of the A exciton of $W Se_2$ [77] by means of the third derivative theory approach [24]. A qualitative agreement could only be achieved under the assumption that mixing between the third derivative of the real and imaginary part of the dielectric constant occurred, otherwise the third derivative spectroscopy did not seem applicable to bound excitonic states.

Some electroreflectance [78, 79] and thermoreflectance [79] measurements have been performed on Bi_2Te_3 and Bi_2Se_3, from which measurements a good deal of new structure was detected and some assignments were made for the detected peaks. Some thermomodulation work on MoS_2 [61, 56] and $MoSe_2$ [44] seems to indicate the essentially derivative nature of thermomodulation techniques.

Photoreflectance measurements on the exciton at the fundamental absorption edge of GaTe [80] yielded the same lineshapes as those obtained by the technique of electroreflectance and were interpreted by photon quenching of a surface electric field.

Finally, wavelength modulation techniques have yielded a large number of new structures in the optical spectra of GaS and GaSe [81].

A few more modulation work references that have not been quoted can be found in references within the given ones.

In any case the present work should not be regarded by any means as a comprehensive review of the field. It is intended to be an exploration of some of the difficulties and possibilities involved in the observation and interpretation of modulation data. Other techniques, besides electromodulation and thermomodulation, have been very sparsely used on layer materials to date, surely as the time passes by more work will be done. It is hoped that the power of modulation techniques will result in a much deeper understanding of the problems involved in interpreting optical spectra of layer materials.

Until now modulation experiments have been mainly confined to the photon energy region below 6 eV. With the appearance of powerful and stable light sources from synchrotron radiation in devoted storage rings, these techniques can be used fruitfully over a much larger photon energy range and certainly very useful information will be obtained.

References

1. M. Cardona: 'Modulation Spectroscopy', *Sol. State Phys. Supp.* **11**, Academic Press (1969).
2. B. O. Seraphin and R. B. Hess: *Phys. Rev. Letters* **15** (1965), 142.
3. W. Franz: *Z. Naturforsch.* **13a** (1958), 484.
4. L. V. Keldysh: *Lov. Phys. JETP* (English Translation) **34** (1958), 788.
5. D. F. Blossey: *Phys. Rev.* **B2** (1970), 3976.
6. J. Dow and D. Redfield: *Phys. Rev.* **B1** (1970), 3358.
7. R. J. Elliott: *Phys. Rev.* **108** (1957), 1384.
8. R. S. Knox: 'Theory of Excitons', *Sol. State Phys. Supp.* **5**, Academic Press (1963).
9. T. Toyazawa: *Progr. Theoret. Phys.* (Kyoto), **20** (1958), 53.
10. F. C. Weinstein, J. D. Dow, and B. Y. Lao: *Phys. Rev.* **B4** (1971), 3502.
11. J. A. Deverin: *Il Nuovo Cimento* **63B** (1969), 1.
12. J. Callaway: *Phys. Rev.* **130** (1963), 549.

13. K. Tharmalingam: *Phys. Rev.* **130** (1963), 2204.
14. D. E. Aspnes: *Phys. Rev.* **147** (1966), 554.
15. Y. Hamakawa, F. A. Germano, and P. Handler: *Phys. Rev.* **167** (1968), 708.
16. S. Koeppen, P. Handler, and S. Jasperson: *Phys. Rev. Letters* **27** (1971), 265.
17. Y. Hamakawa, F. A. Germano, and P. Handler: *J. Phys. Soc. Japan Suppl.* **21** (1966), 11.
18. C. B. Duke and M. E. Alferieff: *Phys. Rev.* **145** (1966), 583.
19. C. M. Penchina, J. K. Pribram, and J. Sak: *Phys. Rev.* **188** (1969), 1240.
20. H. I. Ralph: *J. Phys.* **C1** (1968), 378.
21. D. F. Blossey: *Phys. Rev.* **B3** (1971), 1382.
22. G. Wexler and B. Ricco: private communication.
23. J. Bordas and B. Ricco: to be published.
24. D. E. Aspnes: *Surface Science* **37** (1973), 418; and references therein.
25. J. E. Rowe and D. E. Aspness: *Phys. Rev. Letters* **25** (1970), 162.
26. R. A. Forman and M. Cardona: *II–VI Semiconducting Compounds*, edited by D. G. Thomas, Benjamin, New York 1967, p. 100.
27. Y. Yacoby: *Phys. Rev.* **142** (1966), 445.
28. V. Rehn and D. S. Kyser: *Phys. Rev. Letters* **18** (1967), 848.
29. D. E. Aspnes: reference 1, page 206.
30. J. Bordas: Ph.D. Thesis, University of Cambridge, 1972.
31. S. Nikitine and G. Perny: *C.r. hebd. seanc. Acad. Sci. Paris* **240** (1955), 64.
32. S. Nikitine, J. Schmitt-Burckel, J. Biellmann, and J. Ringeisen: *J. Phys. Chem. Solids* **35** (1964), 951.
33. I. Imai: *J. Phys. Chem. Solids*, **22** (1961), 81.
34. M. R. Tubbs: *Proc. R. Soc. Sec. A* **280** (1964), 566.
35. M. R. Tubbs and A. J. Forty: *J. Phys. Chem. Solids* **26** (1965), 711.
36. S. Brahuis: *Phys. Lett.* **19** (1965), 272.
37. D. L. Greenaway and R. Nitsche: *J. Phys. Chem. Solids* **26** (1965), 1445.
38. D. L. Greenaway and G. Harbeke: *J. Phys. Soc. Japan Suppl.* **21** (1966), 151.
39. CH. Gähwiller and G. Harbeke: *Phys. Rev.* **185** (1969), 1141.
40. G. Baldini and S. Franchi: *Phys. Rev. Lett.* **26** (1971), 503.
41. A. J. Grant and A. D. Yoffe: *Phys. Status Solidi (B)* **43** (1971), K28.
42. G. Harbeke and E. Tosatti: *Phys. Rev. Lett.* **28** (1972), 1567.
43. J. Bordas and E. A. Davis: *Solid State Comm.* **12** (1973), 717.
44. A. Goldberg: (Cavendish Lab., University of Cambridge, England), Private communication.
45. P. I. Perov, L. A. Avdeeva, and M. I. Ellinson: *Sov. Physics – Solid State* **11** (1969), 439.
46. J. Pollman: private communication.
47. J. Bordas and E. A. Davis: *Surface Science* **37** (1973), 828.
48. V. A. Gadzhiev, V. I. Sokolov, V. K. Subashiev, and B. K. Tagiev: *Fiz. Tverdozo Tela (USSR)* **12** (1970), 1350.
49. Y. Suzuki, Y. Hamakawa, H. Kimura, H. Komiya, and S. Ibuki: *J. Phys. Chem. Solids* **31** (1970), 2217.
50. E. Mooser and M. Schluter: *Il Nuovo Cimento* **18** (1973), 164.
51. M. I. Karaman and V. P. Mushinskii: *Fiz. T. Poluprovodrikov (USSR)* **4** (1970), 1143.
52. G. B. Abdullaev, B. G. Tagiev, and V. A. Gadzhiev: *Phys. Stat. Sol.* **B49** (1972), K19.
53. M. Grandolfo, E. Gratton, F. Anufosso Somma, and P. Vecchia: *Phys. Stat. Sol.* **B48** (1971), 729.
54. J. A. Wilson and A. D. Yoffe: *Adv. Phys.* **18** (1969), 193.
55. J. Bordas and E. A. Davis: *Phys. Stat. Sol.* **B60** (1973), 505.
56. J. Bordas: unpublished results.
57. J. L. Jones and G. Brebner: *J. Phys. C. Solid State Physics* **4** (1971), 723.
58. B. L. Evans and P. A. Young: *Proc. Roy. Soc.* **A284** (1965), 402.
59. B. L. Evans and P. A. Young: *Proc. Roy. Soc.* **91** (1967), 475.
60. A. Many, J. Goldstein, and N. B. Grover: *Semiconductor Surfaces*, North-Holland Publishing Company, Amsterdam 1965.
61. G. Weisser: *Surface Science* **37** (1973), 175.
62. A. R. Beal, J. C. Knights, and W. Y. Liang: *J. Phys.* **C5** (1972), 3531.
63. M. Grosmann and S. Nikitine: *Phys. Neudeus Mater* **1** (1963), 277.

64. A. Balzarotti, M. Piacentini, E. Burrattini, and P. Picozzi: *J. Phys. C. Solid State Physics* **4** (1971), L273.
65. Y. Sasaki, C. Hamaguchi, and J. Nakai: *J. Phys. C. Solid State Physics* **5** (1972), L95.
66. B. L. Evans and P. A. Young: *Phys. Stat. Sol.* **25** (1968), 417.
67. R. F. Shaw, W. Y. Liang, and A. D. Yoffe: *J. Non-Crystalline Solids* **4** (1970), 29.
68. R. Z. Allen, M. L. Slade, and A. T. Ward: *Phys. Rev.* **B3** (1971), 4257.
69. A. Balzarotti and M. Grandolfo: *Phys. Rev. Letters* **20** (1968), 9.
70. D. C. Langreth: *Phys. Rev.* **148** (1966), 712.
71. A. Balzarotti, M. Grandolfo, F. Somma, and P. Vecchia: *Phys. Stat. Sol.* **B44** (1971), 713.
72. S. Antoci and L. Mihic: *Solid State Comm. (USA)* **12** (1973), 649.
73. M. Grandolfo, F. Somma, and P. Vecchia: *Phys. Rev.* **B5** (1972), 428.
74. F. Bassani and G. Pastori Parravicini: *Nuovo Cimento* **B50** (1967), 95.
75. H. Kamikura and K. Nakao: *J. Phys. Soc. Japan* **24** (1968), 1313.
76. D. Gili-Tos, M. Grandolfo, and P. Vecchia: *Phys. Rev.* **B7** (1973), 2565.
77. F. Consadori and J. L. Brebner: *Solid State Comm.* **12** (1973), 179.
78. G. Campagnoli, G. Giuliani, A. Gustinetti, and A. Stella: *Surface Science* **37** (1973), 855.
79. A. Balzarotti, E. Burattini, and P. Piccozzi: *Phys. Rev.* **B3** (1971), 1159.
80. K. Taniguchi, A. Moritani, G. Hamaguchi, and J. Nakai: *Surface Science* **37** (1973), 212.
81. E. Burattini, M. Grandolfo, G. Mariutti, and C. Ranghiasci: *Surface Science* **37** (1973), 198.
82. S. Kohn, Y. Petroff, and Y. R. Shen: *Surface Science* **37** (1973), 205.